Label Design and Origination

Repro and pre-press processes

Other Labels & Labeling books:

ENCYCLOPEDIA OF LABEL TECHNOLOGY
Michael Fairley

THE HISTORY OF LABELS
Michael Fairley and Tony White

DIGITAL LABEL AND PACKAGE PRINTING
Michael Fairley

ENVIRONMENTAL PERFORMANCE AND SUSTAINABLE LABELING
Michael Fairley and Danielle Jerschefske

CONVENTIONAL LABEL PRINTING PROCESSES
John Morton and Robert Shimmin

LABEL DESIGN AND ORIGINATION
John Morton and Robert Shimmin

LABEL DISPENSING AND APPLICATION TECHNOLOGY
Michael Fairley

CODES AND CODING TECHNOLOGY
Michael Fairley

LABEL EMBELLISHMENTS AND SPECIAL APPLICATIONS
John Morton and Robert Shimmin

BRAND PROTECTION, SECURITY LABELING AND PACKAGING
Jeremy Plimmer

DIE-CUTTING AND TOOLING
Michael Fairley

MANAGEMENT INFORMATION SYSTEMS AND WORKFLOW AUTOMATION
Michael Fairley

SHRINK SLEEVE TECHNOLOGY
Michael Fairley and Séamus Lafferty

LABEL MARKETS AND APPLICATIONS
John Penhallow

For the latest list please visit: **www.labelsandlabeling.com**

Label Design and Origination

Repro and pre-press processes

John Morton
Robert Shimmin

4impression

Label Design and Origination
Repro and pre-press processes

First edition published 2015 by:
Tarsus Exhibitions & Publishing Ltd

© 2015 Tarsus Exhibitions & Publishing Ltd

Printed by CreateSpace, an Amazon.com company.

ISBN 978-1-910507-03-2

Contents

While every care has been taken to ensure the information, charts, diagrams and illustrations in this publication are correct at the time of publishing it is possible that technology, specifications, markets and applications, or terminology may change at any time, or that the editor's or contributor's research or interpretation may not be regarded as the latest accepted guidance in some parts of the world of labels.

The publishers therefore cannot accept responsibility for any errors of interpretation or for any actions, decisions or practices that readers may take based on the publication content and would advise that the latest industry supplier specifications, standards, legislative requirements, performance guidelines, practices and methodology should always be sought before any investment or implementation is made.

Foreword

The digital revolution has perhaps had its greatest impact on the early stages of a label's life - before it enters the production phases. Processes that traditionally required high levels of skilled manual input have now been transformed by developments in computer software, digital printing and digital workflows.

This book will examine the developments that have taken place in the pre-press arena and will explore in detail the progress of a typical label job from design rough through to its arrival on the printing press. With conventional label printing important manipulations need to be made to the design file in order to anticipate the characteristics of each printing process. These factors will be thoroughly explained.

What will become clear to the reader is that the design to print process is still very complex and that there are many stages to ensuring that the final printed label meets the expectations of the end-user.

The aim of this book is to help provide a comprehensive insight into all these stages, thereby eliminating the opportunities for mistakes that could add unnecessary costs to a design project.

Excellent communication and co-operation between the designer, printer and end-user are an important and essential element in delivering a successful result and supply. Developments in collaborative working models, underpinned by digital workflows, are today increasingly satisfying this requirement.

John Morton and Robert Shimmin
4impression Limited

About the Label Academy

This book is part of the recommended study material for the Label Academy, a global training and certification program for the label industry. The Label Academy was created by the team behind Labels & Labeling magazine and the Labelexpo series of events.

The Academy consists of a series of self-study modules, combining free access to relevant articles and videos with paid text books (both printed and electronic). Once a student has completed a module, there is an opportunity to take an online test and earn a certificate.

It is expected that a Label Academy qualification will become a standard in the industry – for printers/converters, suppliers, brand owners and designers – and assist in providing a benchmark. In addition to its own training, the Label Academy will aim to become a resource provider to the many existing educational programs in the industry. Accredited training courses will be promoted through the Label Academy website and books will be provided at discounted rates.

The Label Academy concept was pioneered by industry expert Mike Fairley. This was in response to a reduction in the number of dedicated printing colleges and the need to standardize training across the world. The label industry also has its own specific training needs – it has some of the widest range of materials, printing processes and finishing solutions of any printing sector.

We are also working with other training experts and authors to ensure that the Label Academy provides up-to-date and relevant training material for the industry.

The Label Academy is supported by the key trade associations, including FINAT, TLMI and the LMAI.

www.label-academy.com

Label Academy sponsors

Thank you to our founding sponsors, without whom this ambitious project would not have been possible:

Cerm

Cerm designs business automation software solutions to meet the specific demands of flexo and digital narrow web printers. Using the latest technology, our team's focus is on innovation and continuous improvement.

Our automation solutions support each step in the printer's integrated workflow – from estimating to production, shipment and data collection – and provide the feature and functionality printers need to gain efficiency and improve profitability.

Cerm inspires collaboration and helps printers remain competitive in the market and deliver the best products possible. We are proud to sponsor the Label Academy and contribute to the future of the narrow web printing industry.

www.cerm.net

Flint Group Narrow Web

Flint Group Narrow Web has the products, the solutions, and the technical experts to handle any print situation. Providing solutions for food packaging, sustainability, increased bottom line, efficiency, and uptime – delivering the basics needed to run a successful operation, and the expertise to go above and beyond to another level of success.

Our experts provide solutions to your printing problems with the innovative products and services that have made us an industry leader around the world. Wherever you are, we are – available to help you reach your business goals today and into the future.

Continuous improvement is paramount to Flint Group; we are proud to sponsor the Label Academy and the benefits it will bring to the future of our industry.

www.flintgrp.com

Gallus Group

The Gallus Group with its production sites in Switzerland and Germany is a leader in the development, production and sale of narrow-web, reel-fed presses designed for label manufacturers. The machine portfolio is augmented by a broad range of screen printing plates (Gallus Screeny), globally decentralized service operations, and a broad offering of printing accessories and replacement parts. The comprehensive portfolio also includes consulting services provided by label experts in all relevant printing and process engineering tasks. The Gallus Group is a member of the Heidelberg Group and employs around 430 people, of whom 253 are based in Switzerland. The group headquarters is in St.Gallen, Switzerland.

www.gallus-group.com

MPS Systems B.V.

Producing high-quality label printing depends on several factors; one of them is the operator of the press.

As a press machine builder since 1996, MPS Systems B.V. knows how important training and education on subjects like pre-press, label printing and finishing is. For label printers, it is critical that their operators keep up with pre-press and press developments in addition to label trends. Therefore, MPS sponsors the Label Academy, to advance operator's passion for printing, share expertise and help multiply benefits.

The MPS slogans of 'Printers First' and 'Technology with Respect' have always underlined the core philosophy of MPS from press design to operator satisfaction. We develop our presses with a strong focus on user-friendliness and respect for the press operator: Printers First.

www.mps4u.com

HP Indigo

HP Indigo is a global leader in digital printing, with a broad portfolio of digital presses and workflow solutions. Indigo's proprietary Liquid Electrophotography (LEP) technology delivers exceptional print quality for the widest variety of applications including labels, flexible packaging, shrink sleeves and folding cartons. HP Indigo's digital presses match gravure print quality satisfying the most demanding brands.

A division of HP Inc.'s Graphics Solutions Business, Indigo serves customers in more than 122 countries, including many of the top label and packaging converters worldwide.

www.hp.com/go/labelsandpackaging

UPM Raflatac

In a little more than three decades, UPM Raflatac has become one of the world's leading manufacturers of pressure sensitive label materials, developing and leveraging the latest innovations in adhesive technology. Our film and paper label stocks are used for product and information labeling across a wide range of end-uses – from pharmaceuticals and security to food and beverage applications.

We are an engineering driven company with industry-leading products known for their consistent high quality and top performance. We are also known for the high performing supply chain and undisputed leadership in the area of sustainability. UPM Raflatac's dedication to innovation, sustainability and top quality is matched only by our commitment to service excellence. We call it the Raflatouch.

www.upmraflatac.com

About the authors

4impression Training

4impression are specialist providers of training across a wide range of print and packaging related subjects. Staffed by industry trained tutors and supported by a network of print and packaging suppliers, the company delivers face to face to courses providing understanding of print processes, embellishments, materials, origination and finishing. Recently 4impression wrote the FINAT Educational Handbook which covers all aspects of self-adhesive label manufacture. They have also produced a comprehensive range of learning resources for the FINAT Knowledge Hub.

As authors of this book 4impression are uniquely positioned to offer additional personalised training to readers who require more insight into its content. The directors of 4impression, colleagues from their days working for the Jarvis Porter Group, are passionate about print and have a long track record in delivering courses to major packaging users and their supply chains.

John Morton

John has hands-on experience of all the major printing processes and has held operational and technical development roles at director level in the packaging sector. A qualified printer, John's career spans magazine production, commercial print, packaging and label production. Before joining 4impression John was actively involved in the Unilever advanced printing and decoration training courses attended by delegates from operations around the globe.

Robert Shimmin

Robert has held senior marketing and business development positions in the print, packaging and label sector spanning more than 20 years. He is a regular contributor of articles to the print and packaging trade press and has supported initiatives that seek to build awareness of the latest research and innovations emanating from UK universities. In addition to his involvement with 4impression he runs Shimmin Associates, a research and marketing consultancy offering support to both UK and international customers in the label and packaging sector.

Paul Jarvis

Paul Jarvis, formerly chairman of Jarvis Porter Group PLC, oversaw its growth to become one of Europe's leading packaging suppliers with a turnover in excess of £100 million, employing 1,600 people in 7 countries including the United States. Paul was a director and founder member of the Leeds Training and Enterprise Council and represented CCL Label on the main board of FINAT (the world-wide association for self-adhesive labels and related products). Paul provides strategic direction to the packaging and print sectors capitalising on his vast experience and global network of contacts.

www.4impression.com

4impression

Chapter 1

—————

Introduction and overview

—————

The aim of this book is to provide a comprehensive overview of what happens to a label or pack design before it arrives at the press ready for printing.

An appreciation of the design to print processes and related terminology is the key to ensuring that each print job meets expectations and that any problems and inconsistencies are anticipated and eliminated from the value chain.

The processes involved in pre-press can be very complex and it is at this stage that costly mistakes can be made.

—————

EVOLUTION OF REPRO AND PRE-PRESS PROCESSES

The term pre-press describes all the activities involved in setting up and preparing packaging and labels for printing. The key components that make up pre-press activities can be seen in the flow chart below (Figure 1.1).

Design and origination – includes all aspects of the label/pack design and the creation of the artwork (finalising and converting a design).

Preparation for printing – features all repro activities including the process of finalising and optimising artwork and color from a production brief.

Proofing – occurs throughout design to print as a means of pre-viewing designs before printing and underpins approvals at each phase.

Output – production of films, plates and other components required for the final print stage.

All these activities will be dealt with in detail in the following chapters.

Components of pre-press

Design/ origination	Preparation for printing	Proofing	Output

Figure 1.1 - The key components of pre-press

1

EARLY PRE-PRESS ACTIVITIES

Traditionally most pre-press operations were based on photographic processes and involved highly skilled manual input.

The following processes and activities were typically part of traditional pre-press operations.

- **Typesetting** – the manual arrangement of text elements (often moveable type) completed by typesetters.
- **Artwork preparation** – the manual creation of artwork onto boards using a combination of images and text.
- **Proof reading** – laborious reading of text and content by a proof reader in order to identify and correct any errors.
- **Copy-editing** – a skilled and manual activity that improved the accuracy, structure and style of content often using a system of annotations (mark up language).
- **Proofing** – the time consuming reproduction of the artwork, often using actual materials and production equipment to be used in the actual production of the job.
- **Screening** – the skilled adjustment of continuous-tone images such as photographs.
- **Separation** – separating the original artwork using a filter for each color (CMYK). This required photographing the target for each color using a large format camera. Also the specifying images or text to be put on the printing plates.
- **Inspection** – viewing of films and transparencies with the aid of an illuminated light table.
- **Retouching** – hand retouching of films and plates to rectify imperfections.
- **Manufacturing** – of plates for printing requiring a high level of manual input.

Today most craft based processes in the pre-press arena have been superseded and have dramatically reduced the need for skilled operators.

A key transition occurred in the mid-1980s. The introduction of the Apple Macintosh and PC along with page makeup software (eg Adobe InDesign, QuarkXpress and the Adobe PostScript* page description language) facilitated the digital manipulation of content on-screen, as well as the output of film and proofs.

It is these developments that are largely responsible for the increase of computer-aided pre-press techniques. By the early 2000s computers became part of the mainstream for pre-press operations with traditional techniques using photographic techniques, dark rooms and light tables overtaken by more efficient digital processes.

More recently, the implementation of page description languages (such as PostScript and Adobe Portable Document Format known as PDF) has provided a standard for document exchange within the industry.

Adobe Postscript – A page description language created by Adobe Systems for defining the content and layout of printed documents in precise detail, so enabling a computer to communicate with a printer.

NEW ERA OF DESIGN

Today the combination of the computer, internet, scanners and imaging units, enables the designer to perform the whole operation from a location almost anywhere in the world.

In addition to assembling graphics and text gathered from a wide variety of sources, due allowance can be made via the computer for the printing process to be used and even the individual press.

Proofing by ink-jet, thermal dye sublimation, etc. provides virtually press quality proofs prior to film or plate making, at any remote location away from the design office. Last minute changes in color or layout may be incorporated easily and re-proofed without delay, often at the customer's site.

By using this modern technology, fast job turnarounds, design flexibility and production oriented layouts are now the norm, with the resulting reduction in costs.

The introduction of the digital printing process is having an important impact on the speed, efficiency and management of the design to print process.

The design to print process overall is however still very complex and costly mistakes can be made, if care is not taken at each stage.

Planning	Creative	Production

Initial request	Create	Capture	Manipulate	Re-work	Approve	Pre-press	On-press

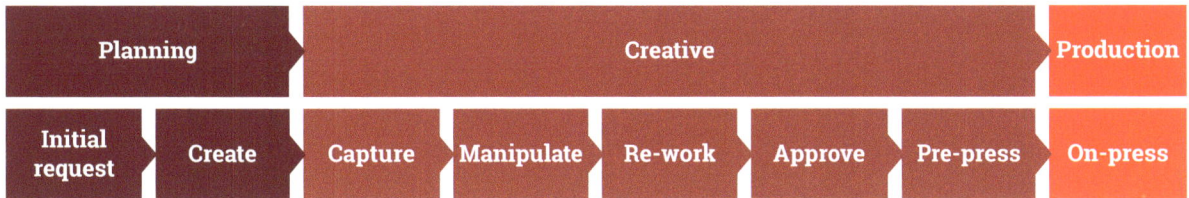

Figure 1.2 - A diagramatic overview of the activities involved in the early planning and creative phases of a project

OVERVIEW OF THE DESIGN PROCESS

Most of a typical design project's costs and time are incurred before it enters production. The planning and creative phases of a job are equally as important as its production, as most of the problems can occur during these early stages (see Figure 1.2).

The fundamental need for pre-press is to anticipate the influence of the printing process to be used and to prepare all data so that its appearance, when printed, meets customer needs.

The above diagram shows how a project evolves through the planning, creative and production phases with the key steps involved.

An important part of pre-press activity is to understand, quantify and document customer needs and expectations. This challenge will be explored in some detail in this book.

SUMMARY

The next page shows a more comprehensive overview of the many steps undertaken in the transition from design to print (see Figure 1.3). This chart shows 35 typical steps in the design to print process. It is important that each of these elements is dealt with in the correct way, as it is quite common for considerable costs to be incurred when one of the elements is incorrect.

A thorough understanding of these steps will go a long way to minimising these errors. Most of these steps will be dealt with in this Module.

At this stage it is important to note that it is the printing process which will ultimately determine the detail of the pre-press procedures adopted.

Typical steps in the design to print process

Design development	Substrate evaluation	Coating requirements
Idea	Absorbency	Product resistance
Roughs	Opacity	Fast drying and curing
First appraisal	Pick strength	Correct slip requirements
Changes to roughs	Surface profile	Correct gloss
Second appraisal	Whiteness	Low odour and taint
First presentation mock up	Coating solubility	Rub protection
Client changes	Gloss	Non yellowing
Second presentation		Smoothness and lay
Design approved		
Applications	**Ink evaluation**	
Range extension/designs		**Preparation to print**
Further presentation	Light fastness	
Artwork drawn	Rub resistance	Step and repeat
Artwork submitted	Hard drying	Films to printer
Artwork changed	Fast setting	Correct print cylinder dimensions
Artwork resubmitted	Adhesion to substrate	Plates made
Remaining artwork done	Varnishability	Preparation of cutters, foiling etc
All artwork submitted	Shade and hue	Press make-ready
All artwork approved	Strength - pigmentation %	Print begins
Artwork to repro	Trapping	Cutting/finishing
Artwork redrawn	Dot reproduction	
Artwork scanned in	Taint and odour	
Manipulations and additions	Toy regulations	
Film output/separations	Stability on substrate	
Chromalin proof		
More manipulations		
Chromalin		
Final retouch		
Proof print		

Figure 1.3 - An overview of the steps involved in the design to print process

Chapter 2

Design and origination

The birth of a label design typically starts in a client's marketing department. At this very early stage packaging concepts are evolved that seek to meet the marketing objectives of the brand. The translation of these concepts into a successful print job will however rely on many factors being brought together in a controlled manner, to deliver the desired result.

How these factors are managed and communicated, is a critical factor in ensuring that each print job meets the expectations of the final customer.

Paying attention to detail in the early stages will reduce the possibility of having to correct costly errors later in the production process.

This chapter explores some of the factors that need to be understood and managed during these early stages of a label or a pack's development.

WORKING WITH DESIGN AGENCIES

A high percentage of new designs emanate from design agencies. Whilst these agencies are very creative, there can be some technical shortcomings and many have limited knowledge of the processes involved in the production of self-adhesive labels. The implication for any project therefore is that a partnership approach is advocated, which includes technical input from the printer.

A summary of typical design agency strengths and weaknesses are listed below.

Design agency strengths
- Visionary
- Enthusiastic
- Perceptive to global trends
- Seek to push printing and conversion technologies to the limit

Design agency weaknesses
- Unawareness of graphic limitations
- Tendency to stray from the technical brief
- Limited knowledge of artwork preparation, repro and print processes

THE INITIAL STAGES OF BRAND/DESIGN DEVELOPMENT

At the outset, design concepts are evolved that take into account a wide range of design and marketing factors. Relevant information from the list below should be factored into a design brief, which will be used internally or by the appointed design agency.

A well prepared design brief will typically include the following;

- **Market brief/market overview/competition** – Market profile data and assessment of the marketing/competitive landscape.
- **Product background** – Essential background on the product, its characteristics and key objectives.

- **Brand guidelines** – including guidance on logo/corporate colors etc.
- **Target consumer** – consumer profile data and target audiences.
- **Mandatory elements** – obligatory packaging elements including legal copy, barcode data, ingredients, warnings.
- **Production requirements** – technical parameters such as web-widths etc. to be established.
- **Product aspirations** – the positioning or required perception of the brand will impact on the label/pack design and should be communicated in the design brief.
- **Costings** – budget parameters/guidance should be established.
- **Timings** – details of critical dates should be provided e.g. market entry/launch date, production dates etc.
- **Die-lines/profile** – usually an important start point for new designs, as it provides the boundary within which the design is established.

- **Maximum number of colors** – there may be technical/cost reasons for establishing an upper limit on color usage.
- **Substrate specification** – early specification of the material to be used may be critical for technical/design reasons.

All the above points must be considered in order to achieve the design objective.

Consultation between all parties, at all stages in a label project is vitally important in agreeing a detailed specification which contains all the relevant information relating to each part of the project.

The specification must be completely adhered to throughout the design to print process and brand owners must ensure that the specification is available to all parties when multi-site manufacturing is involved.

From the design brief the first stage in the design development process is the generation of initial design roughs/concepts (see Figure 2.1).

The second stage is to refine and create visual

Figure 2.1 - Design roughs are evolved that seek to capture the marketing/design brief

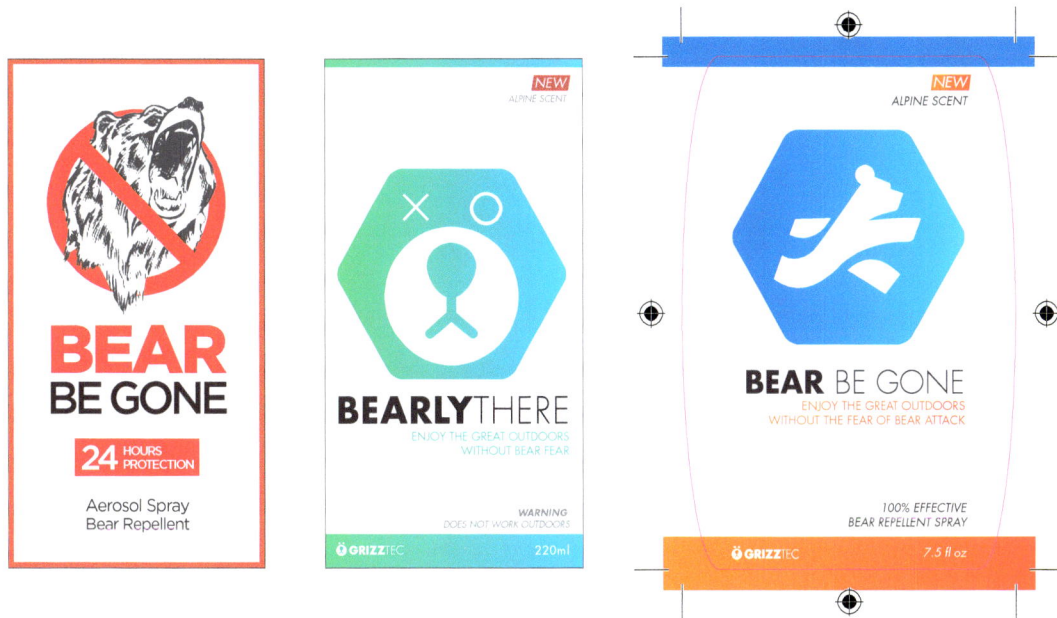

Figure 2.2 - Examples of more fully worked up design visuals

form either for internal discussion or for further market research (Figure 2.2).

AESTHETICS VERSUS PRACTICALITY

Often the aspect of aesthetic design is considered in isolation and sometimes overshadows the technical considerations which must be met to achieve an acceptable end result.

In many cases design creativity must be tempered by the more practical aspects such as cost, technical constraints of the printing process and the many mandatory requirements that need to be accommodated within the design.

Whilst the printer/converter may have little or no input at this stage, he should be consulted regarding the practical aspects of producing the job with his existing facilities, before the creative process gets underway.

THE CONCEPT OF TOTAL APPLIED COST (TAC)

It is important to note that the total cost of packaging

or labeling may have an important impact on the design brief.

The concept of "total applied cost" will often influence the print specification and the decoration/packaging format selected.

The total cost of a label is not merely the cost of the material and its conversion; there are a much wider range of factors to consider.

In-mold and direct decoration, for example, often requires that the user stores large quantities of pre-labeled containers on-site in order to cope with the flows of demand across a number of variants. Storage, inventory and obsolescence are all part of the labeling cost and have to be considered at the very start, when decisions are being taken as to how a product will be produced, presented and marketed. Likewise the cost of application equipment, manning levels and labeling efficiencies are likely to be key components of the cost equation.

The concept of total applied cost encompasses all costs attributed to the labeling process from start to finish. Evaluating different decoration methods using this wider definition of costing can dramatically influence the selection process. Some decoration formats may actually be ruled out on the results of a total cost calculation and this may therefore influence the design brief.

A costing model that takes a total applied cost approach offers a new perspective on the cost of decorating a pack. Total applied costing includes a host of factors such as the cost and efficiency of label application, investment in capital equipment and machinery change parts, logistics and inventory control.

Details of those factors that can be included in a total cost model are provided in Figure 2.3

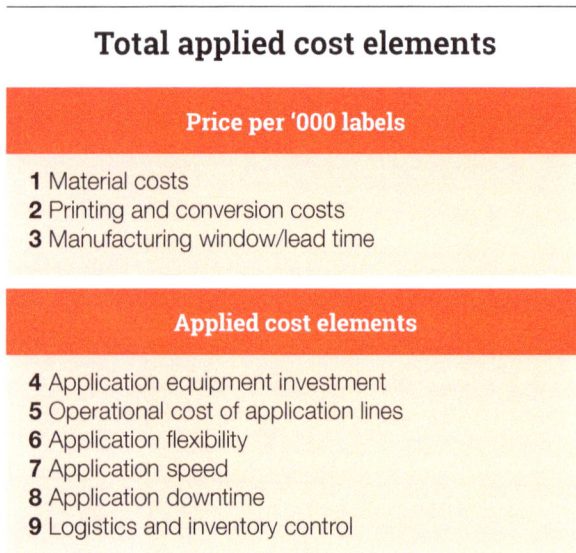

Total applied cost elements

Price per '000 labels

1 Material costs
2 Printing and conversion costs
3 Manufacturing window/lead time

Applied cost elements

4 Application equipment investment
5 Operational cost of application lines
6 Application flexibility
7 Application speed
8 Application downtime
9 Logistics and inventory control

Figure 2.3 - Total applied cost elements

FUNCTIONALITY

Functionality may have an important effect on how a label or pack is produced.

On a self-adhesive label, for example, the function will dictate, to some extent, the materials which will be used, in particular the face material and adhesive. This may limit the type of ink and even the printing process to be used.

An example is scratch-off or temperature sensitive images. These functionalities might be used on a wide variety of substrates, but they both have a specific ink

requirement – i.e. they require a thick ink layer. This factor would then limit the choice of printing process to be used. The scratch-off ink (micro encapsulation) will require it to be printed on a press which controls the printed web without applying abrasive pressure to the printed ink film surface. The temperature responsive inks may restrict drying/curing facilities and the restriction of a maximum temperature in any section of the press. In addition the thickness of the ink deposit may limit type size, dot size and general detail which can be printed.

The most effective route to good design practice begins with a comprehensive understanding of the end-use or final function of the product, and then working back through all the converting, printing and origination procedures, in order to ensure that the desired result is going to be a practical and economic proposition.

A good designer will have at least a basic understanding of the effect of each of the components in the production of a label and how they will impact separately and in conjunction with each other. It is important to consult experts at each stage of production in order to ensure a first time, correct result.

PRINTING PROCESS CONSIDERATIONS

With the dramatic changes in printing technologies over the past few years, choosing the correct printing process to achieve a particular effect is becoming more difficult. The wider use of multi-process presses means that it is possible to combine the strongest attributes of each process in one machine pass to achieve a particular and maybe unique result in an economical way. Offset lithography produces a clear image and fine detail; letterpress a strong color; water-based flexography, high speed and thin ink coverage; UV flexography, controllable ink film thickness and good color coverage; screen printing, high ink film weights and dense color with no show through.

Not to forget the now extensive use of electrophotographic and ink-jet digital technologies which add another dimension to the printing possibilities – including personalisation, adding sequential coding or numbering to the final job.

Finishing processes such as hot or cold foiling,

which can produce a true metallic effect, or glossy or matt varnishing, which can add dramatic effects to the end result, should not be overlooked.

DESIGN EVOLUTION

As a project evolves the design visuals become more developed and can accommodate a wider range of factors including;

- Mandatory elements e.g. warning messages, barcodes, ingredients etc. (see Figure 2.5)
- Die-lines/profiles – will provide the boundary within which the design is to be created
- Maximum number of colors – cost or manufacturing limitations may place an upper limit on the number of colors available to the designer
- Packaging format - Choice of decoration can influence ability to reproduce a design. For example it may not be possible to include certain decorative effects on particular packaging/labeling formats e.g. hot foil stamping is unsuitable for use on shrink sleeve labeling
- Substrates or materials – the choice of materials may impact on design options e.g. metallics

A checklist of critical elements in pack design is provided in Figure 2.4 and will now be explored in greater detail.

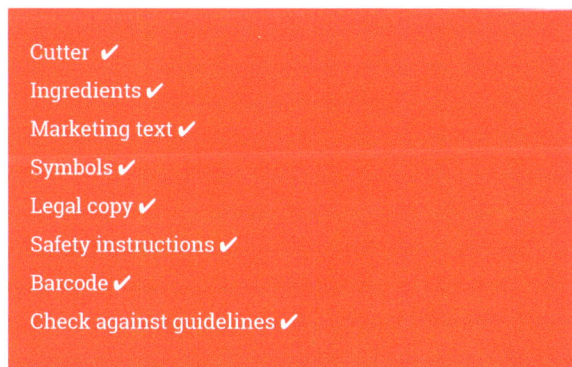

Cutter ✔
Ingredients ✔
Marketing text ✔
Symbols ✔
Legal copy ✔
Safety instructions ✔
Barcode ✔
Check against guidelines ✔

Figure 2.4 - Check list of critical elements in a pack design

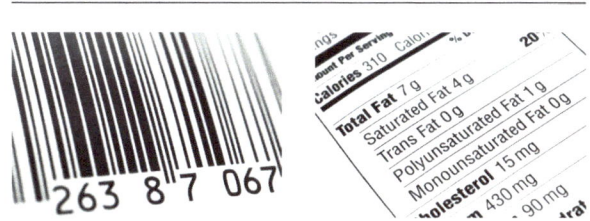

Figure 2.5 - There are numerous legal and mandatory elements that have to be factored into a design such as barcodes and ingredients

MANDATORY ELEMENTS

Mandatory elements that should be factored into the artwork from the outset include – barcodes, ingredients, warning messages, symbols and legal copy.

Compulsory legislation information is the thorn in the side of the designer and can have a dramatic influence on design. More and more information is now mandatory on any type of packaging and labels and the designer is sometimes caught between two opposing requirements - the overall size of the label may be restricted by the size of the package to be labeled, yet legislation might impose a minimum size type to be used to provide certain information. A barcode for example, might have to be printed to certain minimum dimensions. Instructions must be clear and in the language (or languages) required in certain geographical consumer markets. Hazardous products must be marked very clearly and recycling information is also important. The designer may have to work within exactly defined guidelines and yet must include most or all of these requirements, not forgetting the actual brand or product information, which is the main reason for producing the label in the first place.

Sometimes the sheer amount of information required may dictate that an additional label should be included in the final product labeling.

GUIDANCE ON BARCODES

Barcodes should be of high resolution requiring minimal re-scaling. It is important that the barcode printing is in a scannable color. Since most barcode scanners utilise infrared light, avoid using inks with red or orange pigments.

For best results, barcodes should appear on a white background with a no-print area to the left and right of the code.

If printing on filmics, the code must be positioned so that the bars run through the press in the same direction that the film runs through the press, in order to avoid potential image distortion.

COPYRIGHT

Brand images and the company logo may be protected by copyright. This is of concern to the designer and the product manufacturer, but the label printer also has to be aware. If the printer knowingly prints an image on a label that could reasonably be taken to be that of a competitive product, the printer could be held liable.

CUTTER PROFILE OR DIE-LINES

One of the most important considerations in the design brief for a label or pack is its unique shape.

The graphics will be bounded by the outline and of course the design elements will need to be carefully placed within this boundary, for optimum visual impact (see Figure 2.6).

Figure 2.6 - Die-lines/profiles – will provide the critical boundary within which the design is to be created

In the creation of any artwork for packaging design it is vital to start with an accurate template, or cutter die-line.

The die-line clearly defines printable areas, as well as reflecting features such as tear strips, euro slots etc that may need to be taken into consideration when creating artwork. The orientation of text or distortion when a sleeve, for example, is shrunk onto a pack will need to be accommodated within the profile.

Without this information the designer may spend hours laying out the perfect design only to find that it needs to be rearranged to fit a layout with completely different parameters.

Die-lines are typically available from suppliers or can be created in programs such as Adobe® Illustrator.

The die-line may reflect existing cutter stocks available from the converter and will certainly conform to the manufacturing parameters of their equipment.

PRACTICAL CONSIDERATIONS RELATING TO DIE-CUTTING LABELS

The automatic application of the label will run much smoother and quicker if the label profile is kept simple.

There are certain label shapes which can affect the economic running of the press under production conditions.

A few are listed here;

- Acutely sharp corners, star points or any dramatic change to the direction of the cutting edge can cause die-cutting speeds to be reduced. (Figure 2.7)
- Small or thin projections that create an uneven pull on the waste as it is eased away from the face material often tear and remain on the backing liner.

As with any design, aesthetics must be balanced against economic and efficient running of the press. The accuracy of the die and the die-cutting method must be reflected in the complexity of the outline of a label so that penetration of the silicone coating or backing does not occur.

Figure 2.7 - Difficult label profiles - The shaded areas indicate aspects of the label profile that could slow down the press due to difficult stripping. Such areas could be printed to blend in with the container color.

DEVELOPMENTS IN LASER CUTTING AND FINISHING

With the increasing use of laser die-cutting more complex label profiles are now possible, along with more opportunities to create cut-out areas within the label design. The laser uses programs developed from the step-and-repeat function of label origination to guide the laser cutting head around each separate label profile.

Laser etching offers possibilities for personalisation, numbering and coding of labels whereby the images are physically etched into the surface of the material.

NUMBER OF COLORS

All designs should be kept to the maximum number of colors of the converter's printing press that the job will be printed on, including white ink and varnishes.

It is recommended that a color legend is supplied with proofs generated and colors used in a design are best labeled clearly in files.

ECONOMICS

Cost is usually one of the most important factors to be considered in the production of a new label design. The aesthetics of the final design must be balanced against the most economic production method.

The number of colors to be used in the final design will have a huge impact on the cost of production. Many modern label presses, have up to ten color stations, which means that a four color process job can be run in addition to a house or corporate color, plus varnishing and finishing all in one pass.

The printing of very fine detail with small type faces and fine screen rulings can often slow down the running speed of the press and complex cutting, punching and perforating means more units for the press operator to supervise.

Designs incorporating accurate grips between colors or with very fine key lines will also have a negative impact on press speed.

Attention to the detail mentioned here could reduce production costs and if managed correctly may not detract from the overall design.

ARTWORK

The term artwork is defined as the original design, drawings, pictures and text produced by the designer or artist. It comprises all elements of design from which the black-and-white origination and printing plates are made. The process involves the production of finished material suitable for reproduction by any printing method or media. This may be presented as a black and white art sheet, with color overlays, or in disk or CD format, or even transmitted electronically for computer printout.

Once a design has been conceived and created it progresses to finished artwork and may be further amended before final approval is sought. It is also important that the artwork is subject to rigorous checks to ensure that it conforms to specification before proceeding (See Figure 2.8).

FINALISING ARTWORK AND GRAPHIC CONTENT

Following on from the design brief other information is generally required to finalise the artwork. This information will typically include the following components; See Figure 2.9.

Design	Artwork creation	Artwork approval	Artwork measure	Repro

Figure 2.8 - Summary of the artwork process steps from design to repro

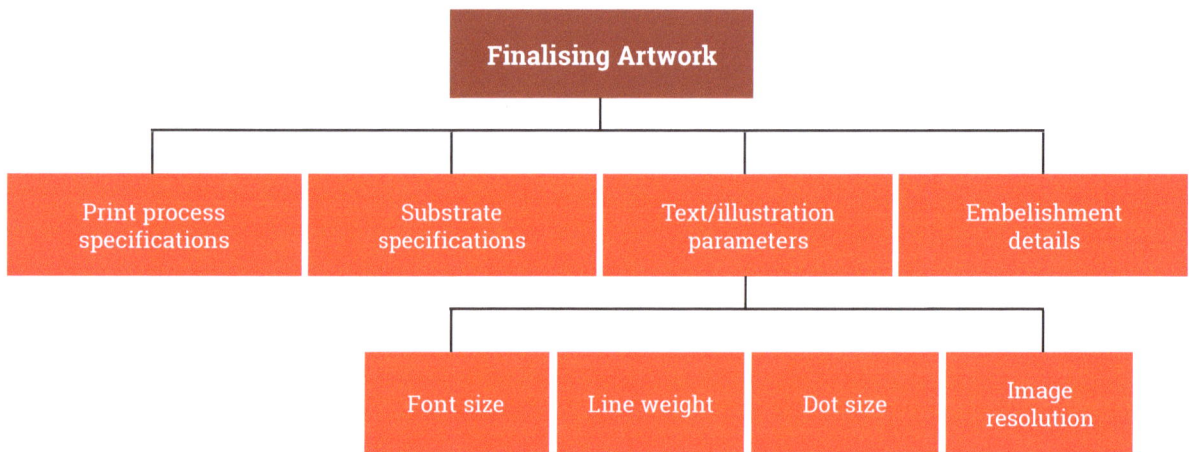

Finalising Artwork

Print process specifications	Substrate specifications	Text/illustration parameters	Embelishment details

Font size	Line weight	Dot size	Image resolution

Figure 2.9 - Components Required to Finalise Artwork

PRINT PROCESS SPECIFICATIONS

At this stage the print processes to be used in the final job are specified. In some cases a combination of printing processes may be required to achieve the desired result.

USING SPOT COLORS

Using CMYK can have its limitations when it comes to color reproduction. If more vibrant colors or an exact color match is required (e.g. for consistent company branding) then spot colors/PMS colors should be identified in the artwork.

USING BLACKS - BLACK VERSUS RICH BLACK

When printing with black color, there are two types of black that can be used.

Black – 100 K: can be used for body copy and barcodes

Rich black – 40 C 40 M 40 Y 100 K: is

recommended when printing blocks of black

Rich black specifications may differ from printer to printer, so it is important to consult with your printer for their advice.

Telling the difference when preparing files on a monitor screen can be difficult since PC screens show richer colors in RGB (red,green and blue). Therefore, it is recommended to get a press proof when printing blocks of black.

The difference between rich black and black is illustrated in Figure 2.10.

Figure 2.10 - The visual difference of a single Black versus Rich Black created from CMYK

```
                          ┌─────────────────────────┐
                          │    Paper substrates     │
                          └─────────────────────────┘
```

Uncoated	Coated and surface treated

| Machine finished | Vellum* | | Matt coated | Gloss coated | Machine glazed** |

| Machine coated*** | Cast coated**** |

```
                          ┌─────────────────────────┐
                          │    Filmic substrates    │
                          └─────────────────────────┘
```

| Polyethylene (PE) | Polypropylene (PP) | Polyester (PET) | Single layer***** | Multi-layer****** |

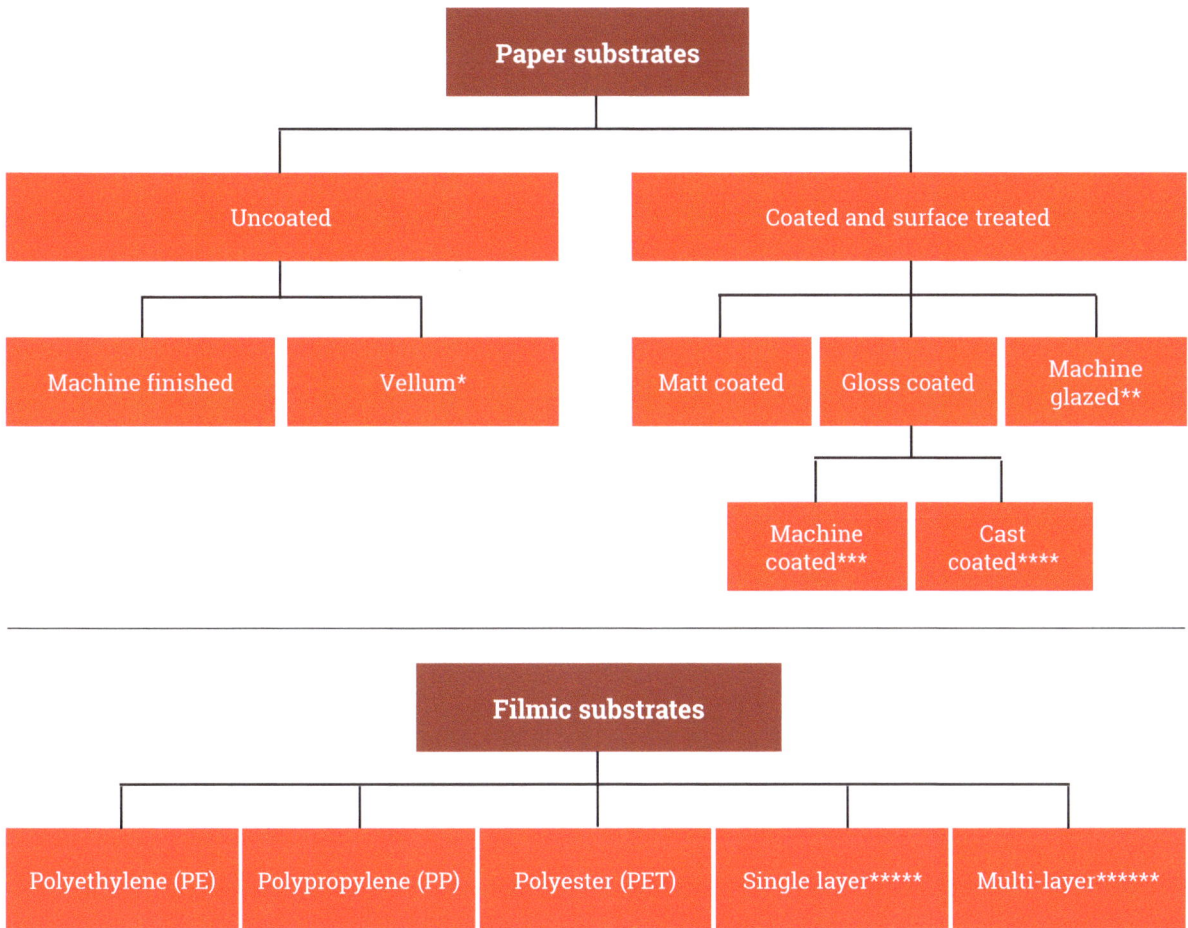

Figure 2.11 - The most commonly used materials

SUBSTRATE SPECIFICATION

At this stage the exact material to be used will be specified along with the supplier details, material grammage/caliper* and the adhesive.

In label printing there is a wide choice of materials available. The more commonly used materials are listed in Figure 2.11;

*Caliper or thickness of a paper or film, measured by a dead weight dial micrometer, usually expressed in thousandths of an inch (mils or thou) in the USA, or in one millionth of a meter (microns) in countries using the metric system.

Vellum* - A strong, tough paper with a high quality appearance, originally made to imitate the fine smooth finish of parchment made from animal skin. No longer noted for its strength, Vellum has become the generic term for very smooth uncoated wood-free paper utilised by label manufacturers for line and solid printing or thermal transfer overprinting.

Machine glazed** – smooth surface achieved using a highly polished steam heated cylinder (not very common)

Machine coated*** – paper coated on the paper making machine

Cast coated ****– clay coating with a high gloss finish

FILMICS

There is an increasing demand for label characteristics that are outside the scope of paper substrates.

Self-adhesive labeling can utilise a wide range of lightweight filmic (plastic) facestocks as shown above in Figure 2.11.

Single layer film***** - mono-layer film

Multi-layer film****** - Films of more than one layer produced extruded with difference performance characteristics in each layer eg printability, dispensability, flexibility, squeezability, stiffness etc

In addition to these filmic materials there is a wide range of other speciality facestocks available including paper/foil laminates, metallised papers/films, synthetic papers.

EMBELLISHMENT DETAILS

Details of surface embellishments for the label or pack are required at this stage and the decoration areas should be specified within the artwork.

Typical decorative effects that can be used include;

- **Embossing -** The process of raising a design or image above the label surface, often through the use of a set of matched male and female dies.
- **Varnishes -** A thin, clear, transparent ink that contains no coloring pigments or dyes. When printed or coated over the top of a substrate and/or printed matter, the varnish provides a protective finish that enhances appearance and increases durability. Varnishes may be glossy or matt.

    If a varnish is to be used, the image or text that requires varnishing should be identified in

the artwork. Typically a spot color named "varnish/spot" is created within the design file.

- **Lamination -** A clear plastic film applied to a sheet or web of labels by heat or adhesive to provide and enhanced, glossy or matt, appearance or for protection.
- **Foil stamping -** Lacquered aluminium foil placed adhesive down on substrate. A heated patterned die is pressed onto the foil to activate adhesive and transfer the image.

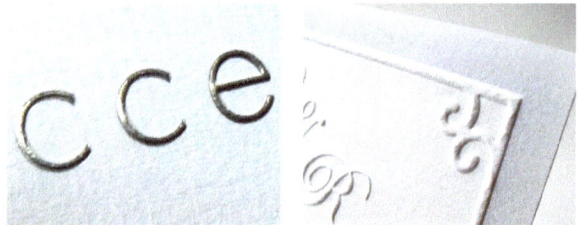

Figure 2.12 - Typical embellishments used in label printing (foil stamping and embossing)

If a varnish, foil or emboss is to be used, the image or text that requires the embellishment should be identified within the artwork. Typically a spot color created within the design file and labeled with the appropriate embellishment description can be used for this purpose.

CUTTER PROFILES

As discussed earlier cutter profiles are typically established at the early stages of design development and provide the boundary within which the design is created.

PARAMETERS ON FONTS AND TEXT

There are a number of factors to be considered when finalising fonts and type as follows;

POSITIVE TEXT SIZING

Minimum type size for positive text is generally 4 point. Type below this point size may not be legible when printed. For best results, small text should be created from one solid color. Screened text can be difficult to read, and slight mis-register on press can affect the legibility of text that is created using more than one color.

Minimum type sizes for particular applications (food or drug labels) are commonly found in the relevant labeling regulations.

REVERSE TEXT

Minimum type size for light-colored text that reverses out of a dark-colored background is 6 point. Type below this point size may fill in and not be legible when printed. Light-style fonts or serifed fonts for reversed-out text are not recommended, as the thinner elements of the letters will have a tendency to fill in.

Type should never reverse out of more than one color and it is recommended that a solid, single-color keyline is used to outline light-colored text.

Printing reversed out text should be avoided below 6 point and the text should be printed directly onto the color (ie not reversed out). See Figure 2.13.

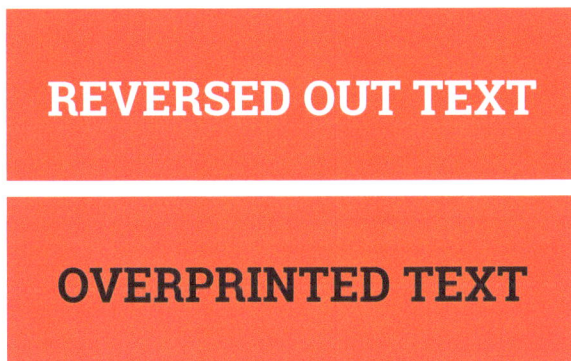

REVERSED OUT TEXT

OVERPRINTED TEXT

Figure 2.13 - Very small text and fine lines need to overprint if they are in a darker color than their background

DROP SHADOW
DROP SHADOW

Figure 2.14 - Drop shadow illustrated

DROP SHADOWS

The use of drop shadows, particularly on small reverse text, is not recommended (Figure 2.14).

The use of a drop shadow introduces an extra color to the background which would have to be printed in perfect register in order to replicate the shape of the letters.

TEXT CREATION IN DESIGN SOFTWARE

Text should always be created in a vector format in design packages. Text created in Adobe® Photoshop® for example, or any other raster-based program, will have jagged, rastered edges, making smaller text particularly difficult to read. Vector based graphics and text will have smooth edges and create a more pleasing result. See Figure 2.15

RASTER VERSUS VECTOR FORMATS EXPLAINED

Like a photograph raster images are made up of pixels with each piece of visual information represented as a small dot that is set in a specific color.

Vector images on the other hand are not made

Figure 2.15 - Rasterized characters from very low resolution to high resolution, compared to a vectorised character

up of dots at all - they are drawings of lines that are represented in the file as mathematical descriptions.

Common file formats for raster images are TIFF, JPG, or GIF.

Common vector file formats are EPS (Encapsulated PostScript), PNG (portable network graphic) and WMF (Windows Meta File).

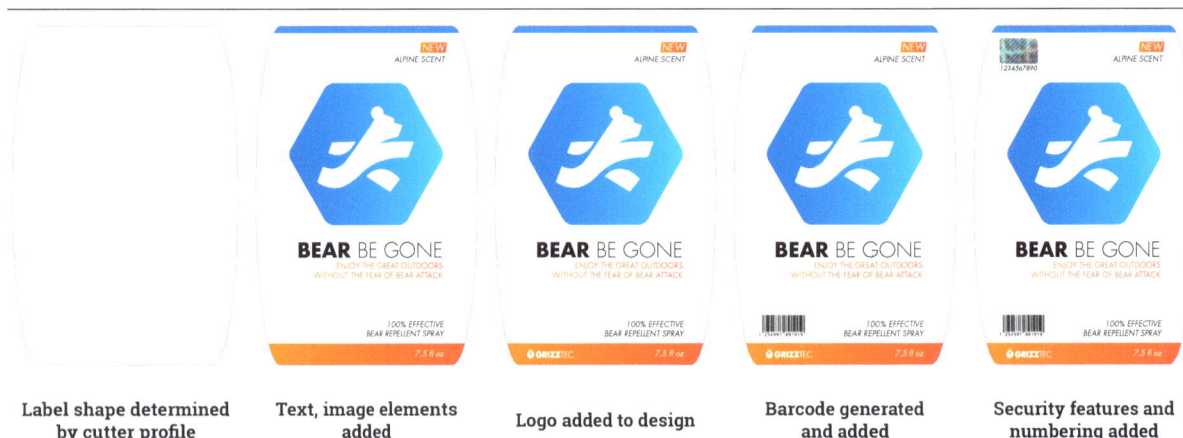

| Label shape determined by cutter profile | Text, image elements added | Logo added to design | Barcode generated and added | Security features and numbering added |

Figure 2.16 - Key label design elements

SUMMARY OF KEY LABEL DESIGN ELEMENTS

Figure 2.16 summarises how all the design elements involved in the origination of a label come to together.

APPROVAL PROCESS

Throughout the development of a pack or label design there are a number of reasons why changes to artwork may take place.

Potential factors resulting in artwork changes can be summarised as follows;

- Change to product specification – re-definitions to the product specification.
- Formulation changes - changes to the product formulation or ingredients.
- Language interpretation – translation errors or clarifications.
- Non-adherence to approval process – issues and errors caused when agreed procedures are bypassed.
- QC checks to artwork content that uncover earlier errors.
- Checks against artwork checklists that identify elements that are missing or incorrect.
- Outstanding information not considered earlier may need to be added now.

Artwork approval by technical, legal, marketing departments or perhaps by the supplier may result in further amendments being required.

After changes to artwork are made, the artwork approval process will need to be conducted again.

PROOFING

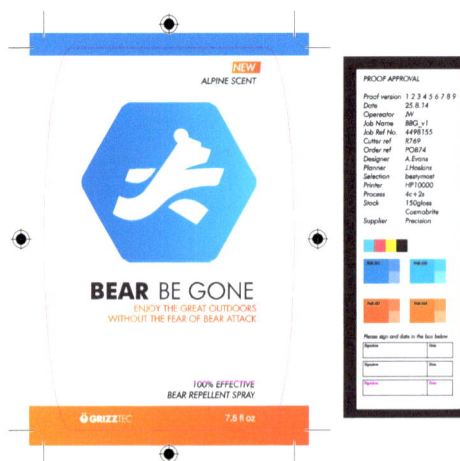

Figure 2.17 - Proofs are typically required by technical, legal, marketing and supplier. At this stage they are used to check copy and layout (but not color)

Design data	CAD driven Laser/resin Container/mold	Substrate Decoration Digital engine	Embellishments	Digital laser die-cutting

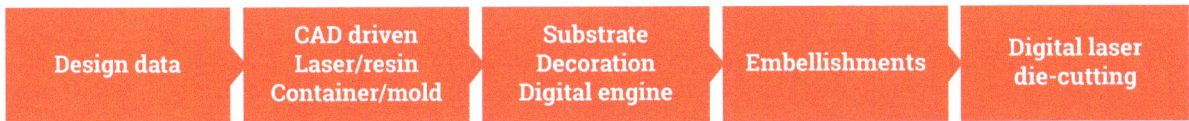

Figure 2.18 - Rapid prototyping described

At the artwork approval stage a soft proof or hard copy digital proof is sufficient to allow interested parties to visualise and make alterations to artwork.

Both these types of proof are termed off-press proofs and are a cost-effective way of providing a visual copy without the expense of creating an actual press proof.

Approved and signed off artwork is now ready to proceed to repro.

PROTOTYPING

Prototyping is important to the packaging market as it allows for the visualisation of products in 3D.

Prototypes are invaluable for retail visualisation and they permit limited-scale test marketing prior to full-scale production.

Figure 2.19 - A realistic 3D mock-up of a label design on screen. *Source: Esko*

In the packaging sector short-run prototypes have typically been produced using the same equipment that is used for full-scale production. This however, is an expensive process and it cannot accommodate multiple versions or last minute design changes.

There are a number of prototyping formats on offer.

VIRTUAL PROTOTYPING

A number of suppliers from the pre-press environment have launched systems and services that use digital technology that permits the 3D visualisation of packaging in a virtual retail environment. See Figure 2.19.

PHYSICAL PROTOTYPING

There are many instances where physical examples of a pack are required, perhaps for test marketing. The main requirement for a physical prototype is that it should look and feel like a professionally produced pack.

The emergence of digital printing has delivered significant benefits to the prototyping process as it eliminates the up-front plate production and make-ready costs.

Color management systems are now available that enable color accurate mock-ups on production substrates to be created.

Inkjet technology is commonly used for the printing of prototypes allowing a finished mock-up to be created within a few hours, without stopping a press.

RAPID PROTOTYPING

The term rapid prototyping (RP) refers to a class of technologies that can automatically construct physical models from Computer-Aided Design (CAD) data.

RAPID PROTOTYPING EXPLAINED

Modern rapid prototyping systems for packaging visualisation take design data in order to recreate a 3D model. The digital printing systems generate graphics which are then applied to the object to complete the prototype.

SUBTRACTIVE 3D PROTOTYPING

Systems that use modelling software linked to a CNC (computer numerical control) milling machine is a method often used to create physical prototypes. The process is called subtractive, in that material is removed (subtracted) from a block to create the final model (Figure 2.18).

Compared to a 3D printer, subtractive rapid prototyping machines deliver smooth surface finishes without needing post finishing.

A CAD model is simply exported to drive the milling machine.

3D PRINTING

The introduction of 3D printing, perhaps more correctly called 'additive' printing is the latest technology that is enabling packaging designers to take 2D designs and extend them to a three-dimensional format. This method of prototyping is called additive in that material is added to construct the 3D model.

3D printing is based on traditional inkjet printing with shapes constructed layer by layer, in fine droplets using liquid photopolymers and then cured by UV light. Packaging design concepts can be brought to life in a matter of hours, but post finishing maybe required to the model.

Labels can be a cost-effective and convenient way to get graphics onto many different packaging formats. Digital printing in particular, is a low cost way to add branding to a physical prototype which can then be used for consumer feedback and test marketing.

Chapter 3

Preparation for printing

Repro is the process of finalising the artwork and producing working files from the design to the production specification.

A key component of the repro process is to finalise all color from the artwork supplied and there is particular reference during the repro phase to critical factors such as color separation, image quality, dot gain and image overlap (known as 'trapping').

The repro process will ensure that the reproduction of the original label design, including the correct color meets client expectations, within the tolerances of the chosen print process.

The traditional method of origination was historically very labour intensive, requiring highly developed skills and was very expensive to undertake.

Today almost all repro is carried out using digital procedures which shorten the production chain and permits greater creativity, with a shorter lead-time.

TYPICAL REPRO WORKFLOW
Typical components of a modern repro workflow are illustrated in Figure 3.1.

FILE STORAGE/SERVER
At the core of a typical repro system is a file storage system usually in the form of a server.

A server is a computer which has been set up specifically to store and 'serve' files to other computer users. In a design studio or pre-press environment, this is essential to prevent the duplication of files and to ensure that an effective backup policy can be implemented.

PDF WORKFLOW
All the components of a repro system are typically linked by a PDF workflow.

Adobe's Portable Document Format (PDF) was developed as a data format for the paperless office, but has established a place in the pre-press environment. Unlike other digital document formats, there is no need for separate text files, fonts, image files or vector art. Acrobat can incorporate them all into the file during PDF generation so that the PDF can be used as a "virtual job bag" to which all necessary pre-press ingredients can be added. A key feature of pre-press workflow systems is the initiation of all processing steps without significant operator involvement.

SCANNERS
Most repro workflows feature a high resolution pre-press scanner that uses a high-speed rotating glass drum (known as a drum scanner). These scanners are used to scan transparencies and photographic images.

Drum scanners give a much more detailed reproduction of an original than flatbed desktop

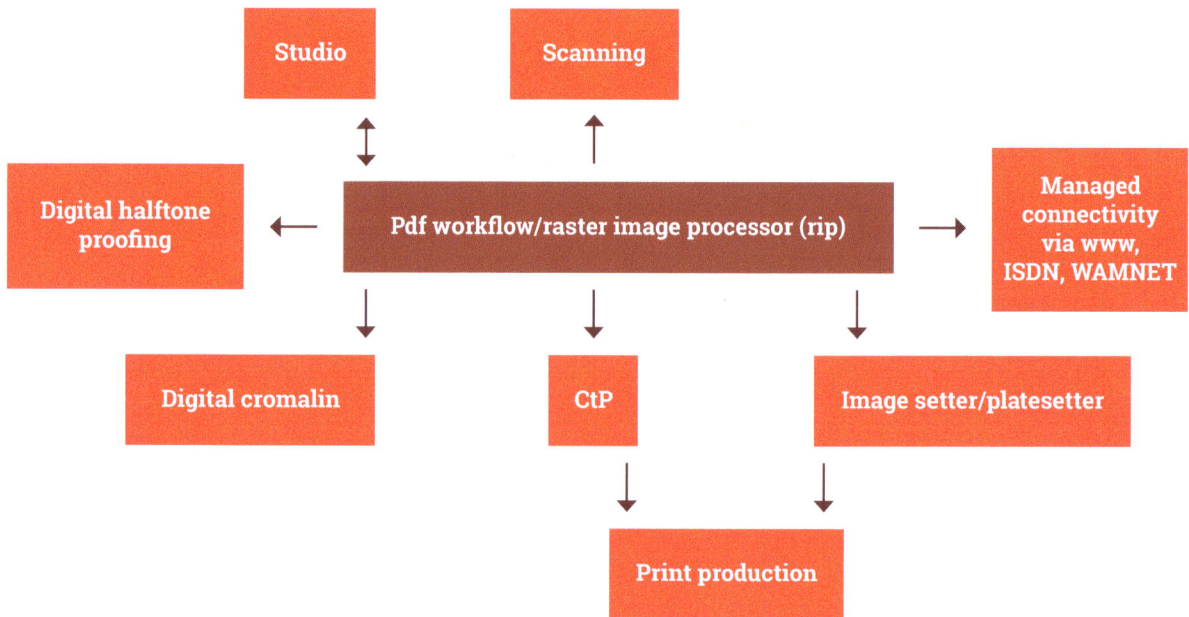

```
        ┌──────────┐              ┌──────────┐
        │  Studio  │              │ Scanning │
        └──────────┘              └──────────┘
             ↕                         ↑
┌──────────────┐      ┌─────────────────────────────────────┐      ┌──────────────┐
│Digital halftone│ ←  │ Pdf workflow/raster image processor  │  →   │   Managed    │
│   proofing     │    │              (rip)                   │      │ connectivity │
└──────────────┘      └─────────────────────────────────────┘      │ via www,     │
                                                                    │ ISDN, WAMNET │
                                                                    └──────────────┘
        ↓                     ↓                    ↓
┌──────────────┐      ┌──────────┐      ┌───────────────────────┐
│Digital cromalin│    │   CtP    │      │Image setter/platesetter│
└──────────────┘      └──────────┘      └───────────────────────┘
                           ↓                    ↓
                    ┌──────────────────┐
                    │ Print production │
                    └──────────────────┘
```

Figure 3.1 - Components of a typical repro workflow

scanners and can capture a much greater range of tones. The performance gap between some of the top flatbed scanners and drum scanners is however narrowing.

IMAGE SETTER

An image setter is a high resolution output device that can transfer electronic text and graphics directly to film, plates, or photo-sensitive paper. An image setter uses a laser and a dedicated raster image processor (RIP) and is usually PostScript-compatible, to create the film used in computer-based pre-production work. These films are used to create the plates that go on the printing press.

Image setters are rapidly being replaced by CtP (computer to plate) systems.

RASTER IMAGE PROCESSOR (RIP)

A RIP is a hardware or software tool that processes a digital PostScript file and then converts it (i.e. rasterises it) to a printable format.

CtP

CtP (Computer to Plate) refers to the manufacture of a printing plate from a computer using laser imaging. The CtP process uses a photopolymer plate coated with a black (ablation) layer that is sensitive in the infra-red range. After removal of its protective film, a laser images the information onto the black mask. In the flexo process the black layer evaporates where the laser beam hits this layer and the actual photopolymer is laid bare in these areas, for the main exposure to follow. After the main exposure, the plate is conventionally washed out, dried, post-exposed and after-treated.

The use of CtP introduces digital pre-press into the pressroom, thereby replacing conventional step-and-repeat technology. It requires adequate digital color proofing and tighter process control. Front-end software and workflow tools must also be adapted to the CtP environment.

PLATESETTER

A platesetter is a machine that generates plates for a printing press. It is similar in function to an image setter, except that instead of producing film from which the plates are made, the plates are imaged and processed in the platesetter.

MANAGED CONNECTIVITY

Most repro workflows rely on services that can distribute digital content securely (e.g. WAM!NET, ISDN or WWW. * WAM!NET is a leading provider of secure managed data file delivery and data management.)

Using a dedicated service such as WAM!NET the content may be kept completely secure, the status of the delivery can be tracked and both senders and receivers can be notified by email or SMS when packages have been delivered to their destinations.

PROOFING SYSTEMS

All repro systems will have a proofing capability. These systems are capable of producing either hardcopy or electronic facsimiles suitable for early, intermediate or final approval. (Proofing is dealt with in more detail in Chapter 4).

DIGITAL ORIGINATION

The digital production route as illustrated opposite (see Figure 3.2) has eliminated many of the process steps required with conventional production. A comparison of conventional versus digital production, highlighting typical devices used is illustrated in Figure 3.3.

All the elements which make up a design can be presented on a computer monitor. Alterations in sizes, layout, and content can be programed with a minimum of delay and without leaving the desk top. Once the designer is happy with what is on the screen, a printout can be made for submission to the customer. This proof can be dispatched direct to the customer, via a web based system, and any changes communicated back to the repro house, almost immediately. Where existing photography, such as slides, transparencies or photographs are to be incorporated these can be scanned, changes incorporated, colors manipulated, etc. and the final result downloaded to the screen for incorporation

within the overall design.

Computers and computer peripherals handle graphics in two fundamental ways. They either create a list of drawing instructions, or define a two dimensional grid of individual picture elements, or pixels.

Conventional

| Data | Film | Plate | Press |

Computer to plate CtP

| Data | Plate | Press |

Direct Imaging

| Data | Plate |

Digital

| Data | Plate |

Figure 3.2 - Possible routes to press - conventional versus digital

SAMPLED IMAGES

Most input and output devices build up graphics in a series of raster scans. Scanners, display monitors, printers, and image-setters are all raster devices and they sample or output graphic objects as a series of separate pixels.

Each pixel is assigned a color value defining the intensities of its primary color components.

Scanned images are therefore referred to as sampled images and their resolution corresponds to the sampling frequency, or distance between individual samples.

If an image needs to be changed in size and it is not practical to re-scan it, pixels must be added or

Conventional production route

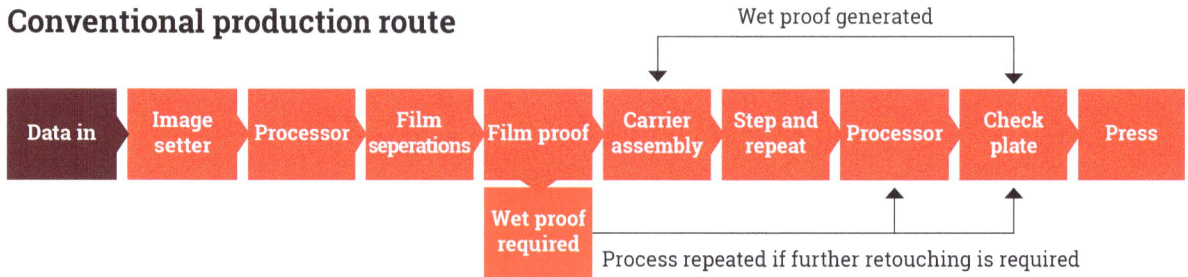

Wet proof generated

| Data in | Image setter | Processor | Film seperations | Film proof | Carrier assembly | Step and repeat | Processor | Check plate | Press |

Wet proof required

Process repeated if further retouching is required

Rationalised digital production route

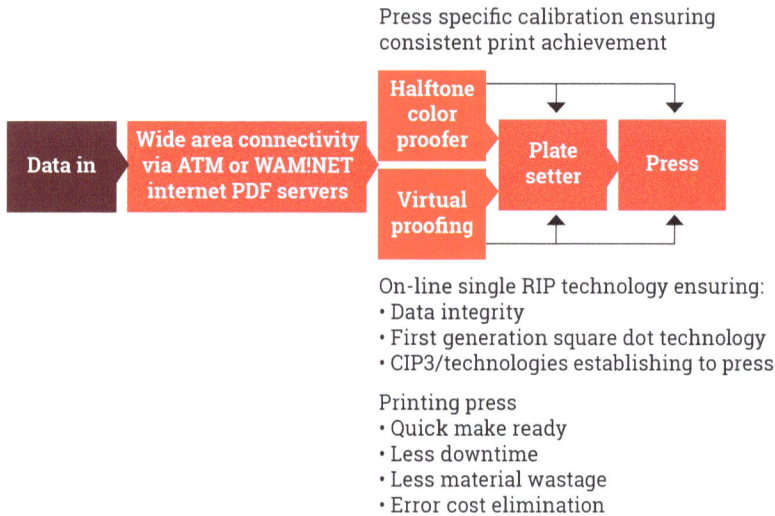

Press specific calibration ensuring consistent print achievement

| Data in | Wide area connectivity via ATM or WAM!NET internet PDF servers | Halftone color proofer / Virtual proofing | Plate setter | Press |

On-line single RIP technology ensuring:
• Data integrity
• First generation square dot technology
• CIP3/technologies establishing to press

Printing press
• Quick make ready
• Less downtime
• Less material wastage
• Error cost elimination

Figure 3.3 - Conventional versus digital production routes highlighting typical devices used

removed in order to maintain the same resolution. This activity is known as re-sampling.

The removal of pixels is a relatively straightforward manipulation, provided that not too many pixels are removed, which would create a 'stepped' effect. This will result in a loss of some detail and parts of the image which will appear out of focus. When the requirement is to enlarge the image the procedure is not quite so straightforward. Pixels have to be added in order to maintain image quality. This can entail adding just one or more pixels beside existing ones. Once an image has been separated into pixels the color range and the tonal dot size is fixed. The color, or grey scale range, is fixed by the bits allowed per

pixel. One bit per pixel will provide either black or white with no grey tones, whilst 8 bits will provide some 256 grades of grey between the black and the white. The first image will have a posterised* appearance whilst the second will have a photographic look.

The same relationship applies when expressing colors. By increasing the bit value the graduation of colors can be taken into many millions. Far above what is in fact reproducible by conventional four color process printing.

The relationship between halftone, and process color screening should relate to the number of pixels in the final size of the image.

Two pixels per dot is the acceptable ratio which will provide even color graduation. If it becomes necessary to adjust the size of an illustration after it has been bitmapped and toned it is better to follow through the procedure of removing the tonal dots, adjusting the pixels and then reapplying the tonal dots in the correct ratio. This will ensure the best possible reproduction.

If the requirement is to re-size the image out of proportion, it is then necessary to add pixels in one direction and remove them in the other. If this is the ultimate intention it might prove beneficial to re-scan the image, oversize, and crop to the desired dimensions prior to creating the pixelised image.

Image processing which involves calculating new values for each pixel can be performed at high speed regardless of the apparent complexity of the image. Any operation which requires the writing of new values for every pixel, such as scaling and rotating, is performed much slower than it would be using vector based methods.

Posterisation- occurs when an image's apparent bit depth has been decreased so much that it has a visual impact (resulting in banding). Posterisation occurs more easily in regions of gradual color transitions.

GRAPHIC FILE FORMATS

Numerous file formats have evolved to describe images for different purposes. They are primarily identified by the degree of resolution, the number of bits that are used to describe each pixel, and the information that can be included in the file header or tag. EPS, JPEG, BMP, GIF and TIFF are the main file formats used for color images. Other formats are also available, but in some instances they do not support four color process requirements.

The PostScript page description language (EPS) allows objects to be incorporated into page files. They can also include a low resolution preview of a bitmap image for display on screen, together with a reference to a high resolution PICT or TIFF file. EPS files can also incorporate the halftone parameters for when the image is output, including screen ruling, angle, dot

shape, and any transfer functions that may have been applied. Some EPS files incorporate a four color image either as a composite, or as a color separated image, in which each color is saved in a separate file.

Tagged Image File Format (TIFF) can be read by most graphics applications. Adobe Photoshop is capable of reading all the extensions that are relevant to graphic arts applications and can convert images from one format to another.

REPRO SOFTWARE AND GUIDANCE ON FILE PREPARATION

Adobe® Illustrator®, Adobe® InDesign® and Quark XPress® files are all excellent programes for typesetting and building multiple page documents, however, they are not always ideal for packaging design. Placed images within these files often have a low-resolution preview, making exact placement of elements a challenge. Support files therefore must be supplied for all placed images, including those that are embedded.

Photoshop® files should be 300 dpi at the size at which they are placed into the final file. It should be recognised that the resolution of a raster file decreases proportionally when enlarged.

When supplying Photoshop® files it is often preferable not to compress the layers in the file. Equally any fonts that are not compatible with the platform being used must be converted to outlines. (Be aware, however, that converting text to outlines, limits the ability to make any content or layout adjustments).

Before sending fonts, it is important to make sure that the licence allows the fonts to be used by both the designer and those outputting the files. It is therefore recommended that one version of the file with the fonts converted to outlines, and one with live text is supplied.

Always include either a PDF or a hard copy proof of the final file so that when the file has been received it can be verified and that there are no issues with fonts or special effects.

If there is a color target that needs to be matched for a CMYK illustration, these also need to be supplied along with swatches of any special match colors.

Dimensions-CAD	Color split/separation	Image resolution (min 300dpi quality)	Press fingerprinting	Screening

Figure 3.4 - A typical repro checklist

DIGITAL PHOTOGRAPHY

Using a scanner, it is possible to incorporate existing illustrations taken with conventional cameras or from a printed document into a file, although this is now rarely the case. The widespread use of digital cameras shortens the chain of procedures.

Digital cameras use a Charge-Coupled Device (CCD), which is a micro-electronic device, to capture the image in pixelised form. The quality of the image varies with the number of pixels that the image is divided into. This will vary with the quality of the camera from as few as 75,000 and as many as 18 mega pixels or more. The images are downloaded to a computer and relayed onto the screen. When a hard copy (print) is required this may be obtained from a printer linked to the computer.

The imaging method of the printer will dictate the quality of the image produced. Once used the CCD device may be cleared and used again and unused images compressed and stored for future reference. It is necessary to check the amount of memory required with that available, as storage, even in compressed form, can use a considerable amount of space.

FILE TRANSMISSION

One of the many advantages of digital origination rests with the ability to transmit illustrations over the internet with no loss of quality. This may be in the form of bespoke Asset Management Systems of FTP (File Transfer Protocol) methods. Once the illustrations are received they may be manipulated and color adjusted to suit the method of reproduction.

A TYPICAL REPRO CHECKLIST

Some of the aspects which must be considered,

irrespective of the repro techniques, are shown in Figure 3.4.

DIMENSIONS CAD

All label and pack dimensions should be checked thoroughly before the repro process commences

Once dimensions are correct then all bleeds should be checked. Bleed is the area to be trimmed off and is a term that refers to printing that goes beyond the edge of the design outline.

Artwork and background colors can extend into the bleed area and this gives the printer a small amount of space to account for movement of the substrate, and any design inconsistencies.

CAD/Dimensions are separated onto a layer of its own. It is important to set CAD dimensions to a color of their own and ensure it is overprinting.

COLOR SPLIT/ SEPARATION

When required for multi-color process printing a full color picture or illustration is normally first divided into four separate subtractive color components by either electronic/laser scanning or through the use of color filters, in a process camera prior to the production of the printing plate.

The term 'color separations' usually refers to the set of four films, one each for the yellow, magenta, cyan and black (CMYK) used for plate production. (Figure 3.5)

Each color on the film is represented as lines of dots set at specific angles which, when overlaid, will combine as layers of dots forming tiny repeat rosette patterns that simulate shades of color when seen at a

Figure 3.5 - Color separations

Figure 3.6 - Illustrates the effect of creating an image from 3 colors instead of 4 - the image on the right has had black replaced with blue. It can create a similar effect to that of the black if these is not much black in the image

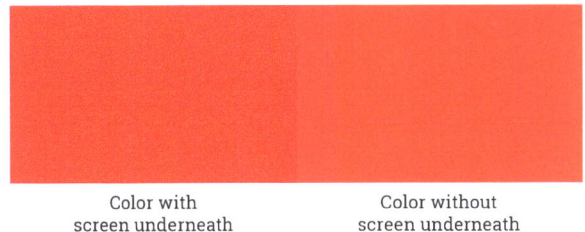

| Color with screen underneath | Color without screen underneath |

Figure 3.7 - The effect on color density when using an additional screen tint

distance.

If an image is required for line printing only, the separated films still represent each of the colors, (which in some printing processes may be more than four colors), but there is no dot formation or rosette pattern involved.

The print-ready data is streamed to a storage medium for incorporation into a finished assembly at a later stage or exposed directly to film or plate.

There may be circumstances where for cost or other reasons an image has to be created out of 3 rather than 4 colors. The visual impact of this is illustrated in Figure 3.6.

COLOR SPLIT CHECK LIST

Below is a check list of factors to be considered when preparing the color elements of a job for repro:

- Determine the number of colors and ensure the palette only contains these colors and that tints are specified correctly
- Minimise the amount of colors where possible without compromising the design/branding
- Detect and redefine colors
- Spot colors in text - There are times for example, in which color text is supposed to be obtained from 4 color process, however this may not be possible due to register problems. If this is the case then the text has to be redefined as a spot color and approved by the customer
- Spot colors in images - if images are required in spot colors, channels need to be generated.

COLOR DENSITY

When printing an image, large areas of flat tone and gradients are often difficult to reproduce. In order to achieve correct color density therefore, it can be beneficial to run a screen tint underneath the solid color.

The effect on color density when using an additional screen tint is illustrated in Figure 3.7.

A typical example when running CMYK would be to introduce a 40% screen tint of cyan underneath a black to increase its density.

TRAPPING

An important aspect of origination is to make sure that it is economically practical for the intended printing process and press to print the image repeatedly, to an agreed quality. The term quality relating to the printed image is wide ranging, but register is a key element and takes the leading position in quality assessment.

With any printing process, misregister or poor print registration is unacceptable.

The output from label printing presses, however superior the engineering, may not be absolutely or consistently precise. Misregister can be caused by mechanical variation within the printing press and also due to temperature changes or movement of the substrate as it progresses through the printing units.

As the printing process is a mechanical process, quality limits should be imposed relative to the complexity of the final result, which would include the economic running speed for the type of label, substrate stability and final application method.

If a label design calls for line colors to butt up to each other (edge to edge) then any movement of the web will cause gaps or unsightly overprinting where the colors meet. The press will not be able to run economically without considering the effect of trapping.

The problem can easily be managed at the repro stage. The basic intent of trapping is to provide an overlay between adjoining colors as a part of the overall design (Figure 3.8). Trapping usually involves expanding the lighter of the two colors to overlap into the darker one.

Computer generated artwork can be trapped by applying a colored stroke or outline to each of the colors affected. Half the width of the stroke will fall inside and half will fall outside the element it is placed against. When a stroke is specified to overprint, the adjoining color will be trapped under the outside half (Figure 3.9)

This technique however needs to be used with caution when text is included. Such trapping used with small or fine type could ruin the image by making the type unreadable.

A number of software programes include automatic trapping functions (often referred to as

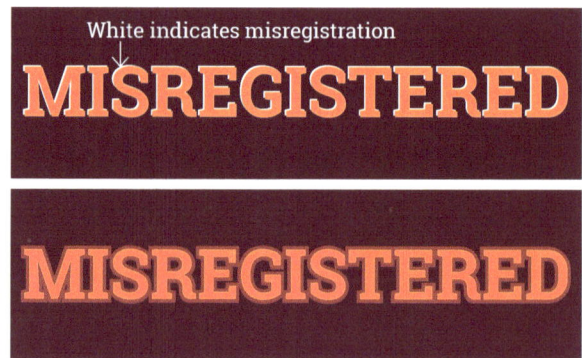

Figure 3.8 - The extra tolerance added is called 'trapping' - Trapping is used to minimise the white halo effects of misregister

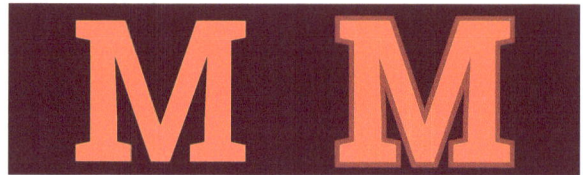

Figure 3.9 - Simple trapping can be done using overprinting strokes. The RH image shows the repro with increased "grip" (overlap) which makes allowance for any movement in the substrate or slight variations in the print register during printing

'spreads' and 'chokes') that can be applied to any area where two colors butt up.

A "choke" is a specific adjustment or distortion of an image whereby the perimeter, in total or in part, is slightly pulled in (choked) towards the center. Choking of an element is normally used in conjunction with the "spreading" of a neighbouring element to guarantee that there are no color fringes or white borders around the image due to misregister.

Another technique for negating the effects of misregister involves the creation of a black key-line to cover the abutting edges and also form an integral part of the line printed elements of the label.

SUITABILITY FOR THE PRINTING PROCESS

In terms of overall printing quality the various processes are all capable of producing printed images on a variety of substrates to a high standard. Each process requires specific considerations when preparing for plate making. One of the greatest benefits to be derived from computer prepared origination is that adjustments can be made in the software programs and applied automatically during preparation of the color separated film. It is even possible to take this a stage further by incorporating the requirements of an individual press should this be desired.

Automatic controls may include such things as minimum line thickness or dot size, dot gain between what appears on the film and what will appear on the substrate. Variations of ink viscosity (ie the thickness of ink being deposited) will affect the color density of the printed image. Such variations may be allowed for by building the necessary information into the program, to be automatically implemented in the origination.

REGISTRATION MARKS

Typically registration marks are included in the design file of a new label job.

Registration marks used in the label industry are positioned on the edges of the web, in the area that is part of the waste matrix. Each color or embellishment (hot foiling and cold foiling) being printed will have its own register mark which is usually in the form of a cross hair positioned within a circle in the form of a target. The register mark image consists of very fine line work (Fig 3.10).

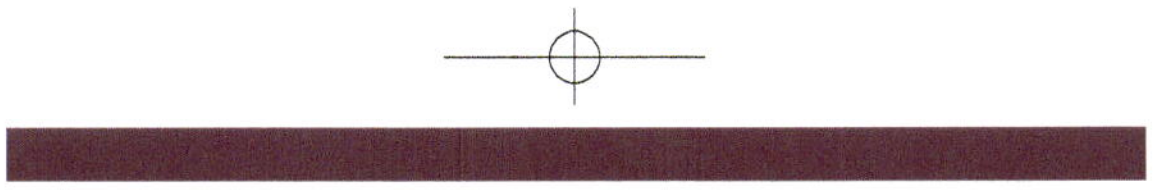

Figure 3.10 - Typical "target" registration mark

Registration marks are a method of monitoring the "print to print" register whilst the press is running. This facility allows the press operator to make manual adjustments to the print register and is particularly useful during the make-ready phase of the job.

Print register is achieved by moving each plate cylinder circumferentially and also sideways until the register mark of each color falls exactly on top of one another. This is called "in-register" and any movement in an individual color will indicate that the print is in "mis-register" (Fig 3.11).

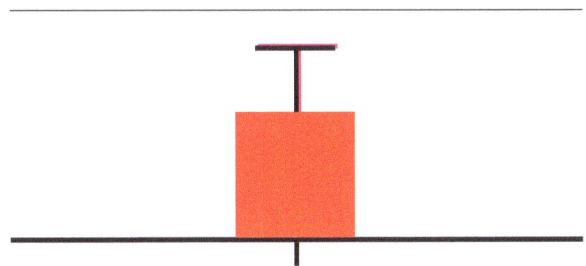

Figure 3.11 - Shows the magenta color out of register

Once the correct print registration has been established and the press speed is increased a register monitoring system will display the register marks as a fixed image on a monitor. This gives the press operator an instant view of the "print to print" register, thereby allowing manual adjustments to be made to the register during the print run.

Another type of registration mark is the one used when the press is fitted with an automatic print registration system. This mark is usually rectangular in shape approximately 5mm x10mm and is printed in the "key" color. This is the color to which the other colors are to print in close register. When the auto system is being used each print unit reads the key color and the system moves the circumferential register into the correct position and then holds it "in-register".

DISPROPORTIONING

More commonly known as "dispro", disproportioning is the calculation that ensures that a printing plate produced in the flat will print an image of the correct size once it is wrapped around a printing cylinder. It involves reducing a plate-ready film in overall image size to compensate for the known distance that the photopolymer plate will stretch or distort on the cylinder.

The amount of distortion will depend upon which print process is being used. The thickness of the printing plate will vary according to the print process and in the case of the flexo and letterpress processes, the thickness of the tape used to mount it on the plate cylinder will need to be factored in as well. Generally, thicker plates and shorter repeat lengths will increase elongation.

Direct-to-plate imaging on to a curved plate, or to a plate already mounted on the cylinder, avoids the need for disproportioning. As the image is applied to a curved surface, no stretching occurs (Figure 3.12).

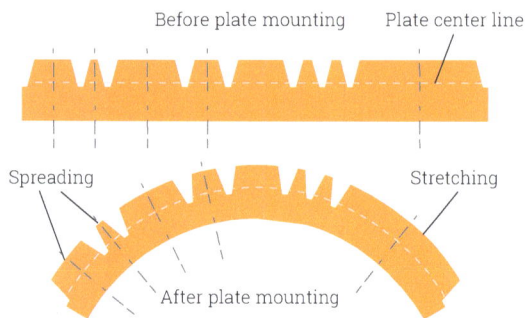

Before plate mounting   Plate center line

Spreading   Stretching

After plate mounting

Figure 3.12 - The "dispro" effect

IMAGE RESOLUTION

It is important to note that in order to achieve a good printed result, a high quality original is a necessity. Generating a high resolution image in Photoshop from a poor quality original will result in poor print quality (Figure 3.13).

When resizing original images a key factor to remember is that the more dots per square inch the higher the resolution. The fewer dots per square inch the lower the resolution. The optimum image resolution for printing is 150 lpi screen = 300 dpi*.

dpi - Dots per square inch expresses the number of lines of halftone dots per square inch. Maybe expressed in-lines (lpi, lines per square inch) or dots per square centimeter.

Effective resolution is the resolution of the images with the effect of scaling taken into account. A 300 dpi image at 400% scaling would be 300dpi x 100 /400 = 75dpi dpi effective resolution (Figs 3.13 and 3.14).

Figure 3.13 - These two illustrations show the impact on resolution on resizing. Increasing the size of the image drastically reduces definition.

IMAGE RESOLUTION CHECKLIST
- Ensure an adequate resolution of original images so that the pixels do not show on print
- Generating a high resolution image in Photoshop from a poor quality original will result in poor print quality
- Optimum image resolution for 150 lpi screen = 300 dpi*
- The more dots per square inch the higher the resolution
- The less dots per square inch the lower the resolution

DOT GAIN

Dot gain is a characteristic of most printing processes where the size of the half tone dot changes, as a result of plate to substrate pressure, during the printing process (Figure 3.15 and 3.16). Because a conventional printing press is mechanical by nature, the dot gain can vary dependent on the age and condition of the press. Dot gain can make a considerable difference to the final printed result and

the highlight and shadow parts of the tonal reproduction scale, giving unacceptable results especially in four color process printing.

The problem with dot gain is that it is only part of the picture. Each press runs at its own color densities. In order to understand the color effect it is important to understand the effect of these densities. (Figure 3.17 and 3.18)

300dpi at 100%

300dpi at 200%

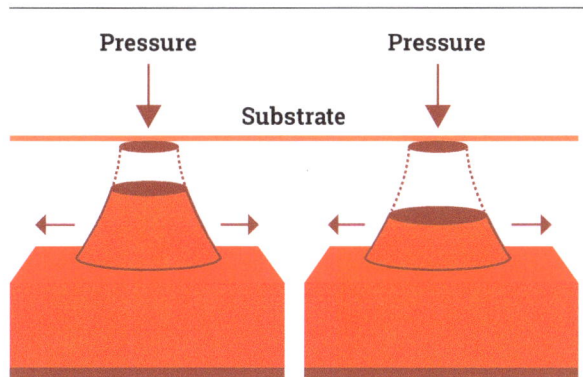

Figure 3.15 - Dot gain - half tone dot changes under pressure

300dpi at 400%

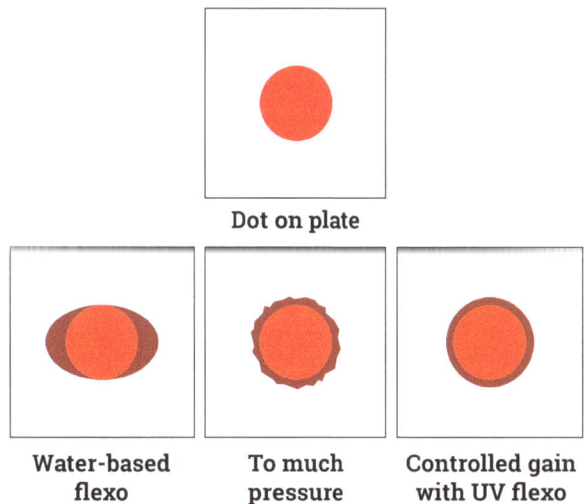

Figure 3.16 - Effects of pressure on dot structure

Figure 3.14 - Image resolution – scaling up reduces the effective resolution of an image

may result in a complete change in the tonal values of color printing. Dot gain is most prevalent and visible in

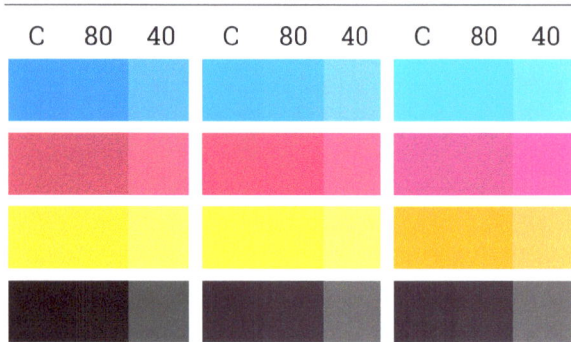

C	80	40	C	80	40	C	80	40

Figure 3.17 - The diagram shows 3 identical dot gains but with 3 different ink colors and densities. They all look slightly different

PRESS FINGERPRINTING

The printing characteristics of an individual press can be measured (via finger printing*) and the results built into the preparation of films and plates to give a predictable printed result.

The established characteristics are only valid if none of the printing parameters change. Any change, be it different plates, inks or substrates requires a new fingerprinting exercise to be undertaken. The same rule applies if the press is modified in any way, or the overall running speed is altered or the anilox changed (in flexo).

Typically a printer will carry out a press fingerprinting exercise in order to obtain, amongst other things, dot gain information. The printed effects of dot gain are illustrated in Figure 3.18 and 3.19.

Original

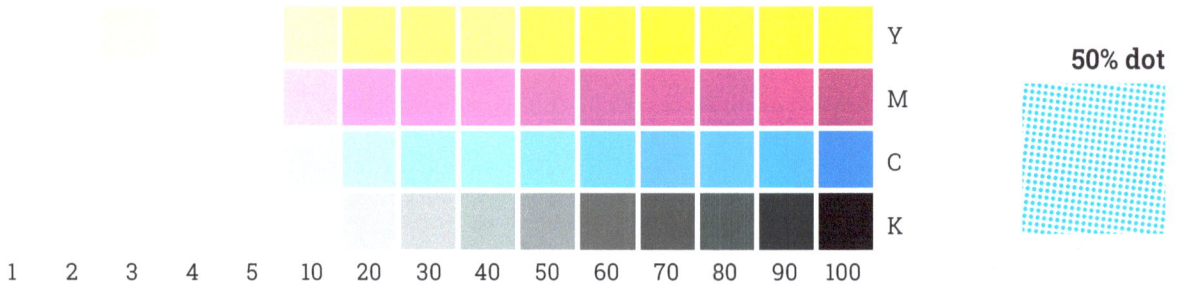

50% dot

When printed

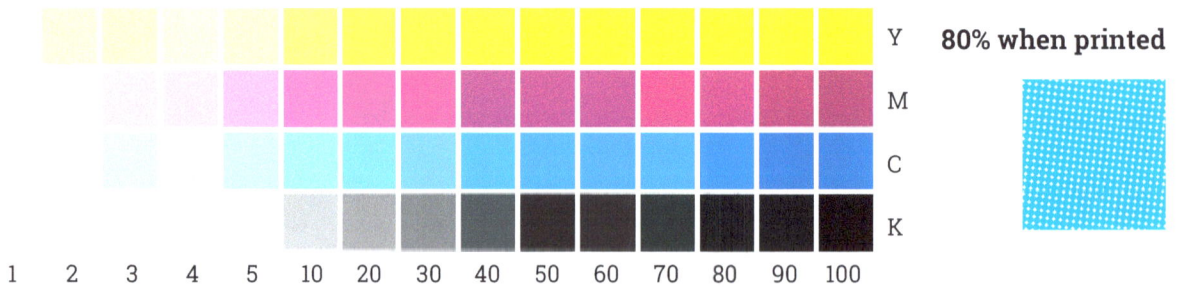

80% when printed

Figure 3.18 - The effects of dot gain

The printing of a special test run on a press so as to determine the registration, dot gain, distortion and other characteristics of the press. Once known, these can be compensated for at the design, film or platemaking stages.

- Dot gain information is required at the repro stage to compensate for a measured gain
- Each press has a unique dot gain percentage
- Fingerprint trials must be individual to each press

SCREENING - SOLIDS VERSUS SCREENING

Tint (screen) – is a solid color which has been reduced in shade by applying a screen during the

50% Halftone

With 15% dot gain

Figure 3.19 - Comparison of images produced with 50% half tone (left image) versus the same image produced with 15% dot gain (right image). To overcome the effects of dot gain on the printed image the repro file has to be digitally adjusted to reduce the size of the dot by 15%.

Figure 3.20 - Cyan separation

Figure 3.21 - 4-color separations

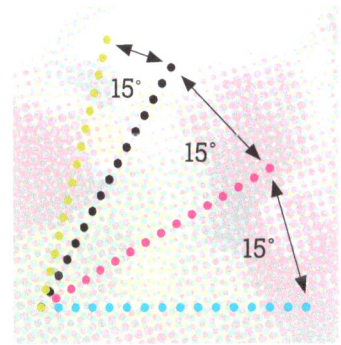

Figure 3.22 - Screen angles

DOT GAIN – A SUMMARY OF KEY POINTS

- Dot gain is critical to color management
- Dot gain is a method of measuring the dot size variation between digital data and press caused by the mechanical process of printing

origination process. The resulting tint is specified as a percentage screen of the original solid color.

Tint (solid) – a dense area of color without screening, usually defined as a percentage of a defined solid color, as opposed to a screened tint of a solid color. A solid tint color is obtained by ink mixing, rather than as a pre-press operation.

Halftone screening is a way of splitting tones into separations that can overprint each other to make different colors (see Figure 3.20 and 3.21). Typically the screens need to be on different angles (Figure 3.22) and a screen's coarseness can vary dependent upon the quality of the material and print process.

ROSETTE

A rosette pattern that is created when all four color halftone screens are placed at the traditional prescribed angles to each other (see Figure 3.23)

Figure 3.23 - Rosette pattern

MOIRÉ

A Moiré pattern, found in both black and white or color halftone printing, describes the irregular

patches - either over the whole image or in certain color combinations (see Figure 3.24). It results from incorrect screen angles being used when overprinting colors or when reproducing from an already printed subject.

STOCHASTIC SCREENING

The use of stochastic screening can avoid the problems of moiré and are ideal for use with Opaltone processes (the Opaltone process is explained below).

Stochastic screening is an alternative to conventional halftone screening in which the image is separated into very fine, randomly-placed microdots (measured in microns), rather than the more conventional grid of geometrically aligned halftone cells. Dots are the same shape and size in most versions, but there is varying spacing between the dots (see Figure 3.25)

Sometimes called frequency modulated or FM screening, stochastic screening eliminates screen angles and the possibility of moiré patterns and provides greater image detail due to the lack of screen rulings and screen angles.

There are no rosette patterns with the process and results are impressive on both coated and uncoated paper and film.

The stochastic screening process involves imaging dots on film using special randomizing software. The software uses mathematical expressions to

Figure 3.24 - Examples of moiré patterns caused by incorrect halftone screen angles. Right-hand illustration shows correct screen angles.

unwanted interference pattern of screen dots caused by combining one, regular, halftone pattern with another similar one, so causing disturbing patterns or

statistically evaluate and randomly distribute pixels under a fixed set of parameters.

Stochastic screening produces smoother

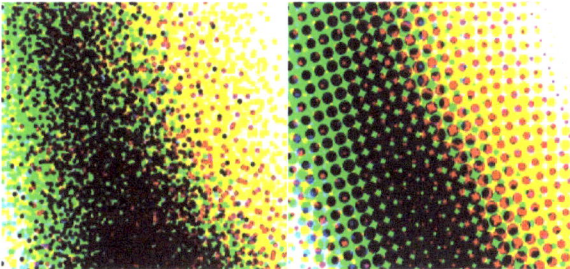

Figure 3.25 - Stochastic screening - Random placing of dots in stochastic screening compared to AM (or conventional) screening.

gradations when vignettes, blends and degrades are involved. It enables almost any combination of colors to be used in the creation of subtle or dramatic effects.

The dot size used in stochastic screening is extremely small when compared to the size of the highlight dot in conventional screening. It is recommended that stochastic screening is used for flexography only after the printer and color separator have undertaken press fingerprinting to determine the ideal dot size and accurate compensation for dot gain.

OPALTONE

Opaltone® is a patented imaging technology that digitally mixes CMY+RGB process inks

The CMYK system can only reproduce a limited color gamut. For example oranges, reds, bright greens and blues are common colors that cannot be faithfully reproduced in CMYK.

The "traditional" way to overcome the CMYK gamut limitations has always been to print expensive spot colors. The Opaltone system is a way of digitally simulating spot colors.

PROOFING

At the latter stages of repro the aim is to help the customer visualise their job as accurately as possible in the form of a proof (Figure 3.26).

Printing proofs are used for checking that all text and graphics and colors come out as expected before going to press. Aside from the finished piece

the proof is often the only part of the production process that the client will see. (see Chapter 4 for a more detailed explanation of Proofing)

Figure 3.26 - Proofs help customer visualise their job as accurately as possible

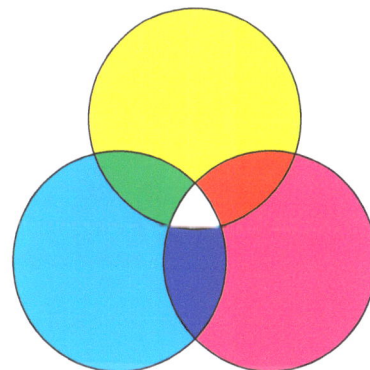

WHITE	=	All three colors
YELLOW	=	Red + Green
MAGENTA	=	Blue + Red
CYAN	=	Blue + Green

Figure 3.27 - The primary colors of light—red, green and blue—as they relate to the additive colors—yellow, cyan and magenta.

ADDITIVE AND SUBTRACTED COLOR EXPLAINED

Color makes up the visible light spectrum, which is made up of red, green and blue (known as the additive colors). Red, green and blue are called additive because when you add them together the result is white light (see Figure 3.27).

Cameras take images and scanners scan in RGB. Computer monitors and web based applications are classed as RGB media. RGB images have much more range of brightness than CMYK images.

Printing on the other hand uses ink and uses the subtractive colors (cyan, magenta and yellow or CMY), plus black (abbreviated to K). CMYK color is also called "process color" or "full color."

The primary colors (hues) used for process color printing subtract light and when overlapped produce other colors and black images. See Figure 3.28.

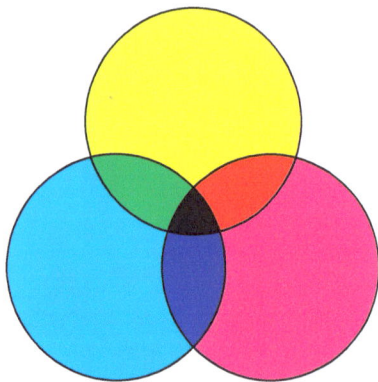

Figure 3.28 - The principles of subtractive color

COLOR CONTROL AND MANAGEMENT

Quite naturally end users expect the color they have seen and approved (via a proof) to be the same as the color they get on the final printed result. All printers will have processes in place (color management systems) to ensure that this is achieved.

Color control and management processes are set out to achieve the following objectives;

- Ensuring color consistency amongst different input and output devices so that printed results

meet expectations
- To ensure that the colors on the monitor are matched by the finished results on the press
- Color matching – the exact matching of a given color sample in terms of hue, value and intensity

Some of the key methods used to ensure good color matching are detailed overleaf;

INK COLOR MATCHING

It is critical in color matching to duplicate the exact hue, value and intensity of the color in the ink blend. Each color should be verified under the correct lighting conditions and the use of appropriate color measurement instruments adopted to ensure an acceptable match. An instrument known as a colorimeter can be used to measure the spectral reflectance of a color and compute numeric values for color intensity, hue and purity.

COLOR CALIBRATION

Color calibration is the process whereby a series of graphic input and output devices are calibrated, using color profiles, in order to ensure color consistency across the design, pre-press and printing operations.

Input devices, such as scanners and digital cameras and output devices, such as monitors, printers, proofing devices and image-setters are calibrated.

The range of colors available to a specific output device, are typically known as the color gamut. The RGB color range is much broader than the CMYK color gamut (which is what most pre-press output devices use). Colors specified using the RGB gamut will often fall out of the gamut range when output on a CMYK device.

ICC FILE STANDARD

The file standard for describing spaces is called the ICC (International Color Consortium) profile. The ICC framework allows pre-press processes to convert color scientifically.

Most artwork packages and operating systems now offer some sort of color management, but the profiles available are generic printer and monitor profiles.

To be completely accurate individual devices used

need to be profiled.

ICC profiles can be used to convert between color spaces maintaining color as close as possible

ICC profile can be used at operating system level or on a document by document basis in Photoshop. See Figure 3.29.

Figure 3.29 - Same data viewed in Photoshop using 3 differet ICC profiles - The impact of ICC profiles demonstrated

New tools for fine tuning CMYK separations are developing which complete the transformation RGB/CMYK to CMYK in a single step.

COLOR BARS

A series of colored shapes printed outside of the finished print area. These bars are often used to verify the accuracy of the printing job and allow the press operator to calibrate the print job and adjust the press if necessary. (see Figure 3.30)

Figure 3.30 - Color control bars help the printer control quality on press

COLOR CORRECTION

Color correction is the process of adjusting an image so as to correct color errors.

Overall color correction is a basic function of color reproduction. This may be required for any of the following reasons:

- Poor scanning
- Poor film/print processing
- Poor color management

All repro houses perform overall color correction and simple aesthetic retouching to varying standards in order to optimise the printed image (see Figs 3.31 and 3.32 below).

Figure 3.31 - When the whites are darker and the blacks are lighter (left image) the image may lack detail. With the correct repro the image appears more vibrant (right image)

Figure 3.32 - Example of color corrected image (before and after color correction)

SPOT COLORS

Spot colors printed by CMYK can be a tricky business as there are limitations on the densities and color that can be achieved. Figure 3.33 illustrates the difficulties in reproducing a spot color using 4 color process and variability that can occur if the same job is produced by two different printers.

The left column of each is the spot color which you can see are identical. The right column of each is the process conversions.

These can vary dependent upon the printer and their relevant dot gains and densities.

If colors cannot be successfully and consistently reproduced using CMYK then single 'spot' colors can be used. This will of course require an extra printing head on the press.

SUBSTRATES CHARACTERISTICS AND EVALUATION

The substrate used will have a distinct influence on the printed result and final color of the job.

There are a range of substrate characteristics that need to be evaluated as part of the specification process.

The image and reproduction will be affected by the following characteristics;

- Absorbency
- Reflectivity
- Color
- Smoothness

Materials act like ink pigments and reflect color to varying degrees. Inks are not 100% opaque (they have an element of transparency) so some of the characteristics of the underlying substrate are transmitted through the inks.

The appearance of an image printed on a white substrate for example will differ considerably from one printed on a brown material. The impact of the underlying substrate on the final result must be

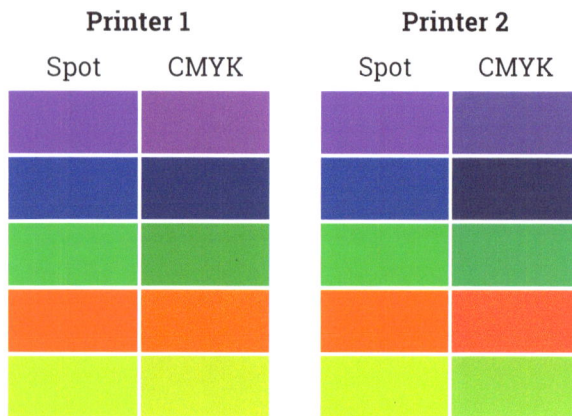

Figure 3.33 - The two panels in the graphic above show two identical spot to process conversions printed by two different printers

Figure 3.34 - The impact of substrate color on the visual appearance of a printed design

carefully considered and anticipated at the repro stage (see Figure 3.34).

OPTICAL PROPERTIES

Other optical properties may have an impact on the final printed result and must therefore be carefully considered. These factors include;

Color – Absorption within the object or product
Gloss – Specular reflection at the front surface of the object
Translucency – Transmission through the object
Texture – Spatial variation in reflection / transmission / absorption caused by surface texture

TRANSPARENCY

It is important to note that inks are not 100% opaque - they have an element of transparency and some of the characteristics of the substrate can often be transmitted to the inks.

If a semi transparent substrate or ink is being used on a pack it is important that the impact of container color or product color is anticipated. For example a dark or strong colored container can have a significant influence on the appearance of the label design. This influence is demonstrated in Figure 3.35 below.

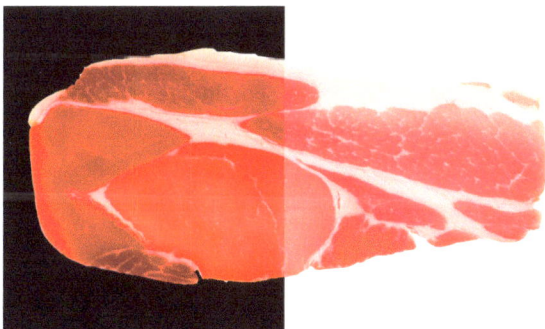

Figure 3.35 - Background influence on color of Containers/Products/Liquids

TRANSLUCENCY

The translucency of an ink or material can also have a dramatic impact on appearance.

INK EVALUATION

The type and characteristics of the inks to be used on a job must be considered and their impact evaluated. Surface coatings to provide additional gloss, surface or product resistance for example, must be evaluated and factored into the specification.

Other factors relating to inks and varnishes to be considered include the following;

- Light fastness
- Rub resistance
- Drying
- Adhesion
- Varnish requirements
- Shade and hue
- Strength/pigmentation
- Taint & odour
- Toy Regulations – the labetling of toys and children's products must comply with the requirements of the relevant national or international safety standards.
- Stability on substrate

All of the above ink considerations will be explained in the Inks, Coatings and Varnishes training module.

PRE-FLIGHTING

Once the repro process is completed the job is ready to go to print. There is however a final process called pre-flighting that is often used in order to confirm that the digital files required for the printing process are all present, valid, correctly formatted, and of the desired type.

The files to be used for the printing of the job are checked to make sure they are in a format that can be interpreted by the RIP (raster image processor). Once the incoming files have passed the pre-flight check, they are ready to be put into production. Without this pre-flight check significant and expensive production delays could result.

PDF is the industry standard file type for submitting data to a RIP. Printers will make sure the pre-flight settings match their specific production requirements.

Files are verified by a pre-flight operator for

completeness and to confirm the incoming materials meet the production requirements. The pre-flight process typically checks for:

- images and graphics embedded by the client have been provided and are accessible
- image files are of formats that the application can process
- image files are not corrupt and are of the correct resolution and color format
- required color profiles are included
- fonts are accessible and compatible to the system and are not corrupt
- confirm that the correct separations are being output

Advanced pre-flight steps might also involve the following;

- converting fonts to paths
- removing hidden objects (i.e. objects outside the printable area and objects on layers below)
- flattening transparent objects into a single opaque object
- gathering embedded images and graphic files to one location accessible to the system
- compressing files into an archive format

ON PRESS CHECK

The on press check takes place after a printing press is set up, but before the print run commences.

While errors should have been corrected during the approval and proofing stages, the main purpose of a press check is to make sure that the color on press comes as close as possible to the agreed color proof.

There can be inherent differences between some color proofing methods (apart from wet proofs) and the printing process itself.

Areas that are commonly evaluated at a press check are as follows;

- Flesh tones or corporate logo match colors
- Overall color balance across the web
- Substrate (checking for correct color, weight or texture).
- Content (looking for missing elements and confirming copy changes).
- Registration (checking sharpness, color overlapping, edges of images and screened type)
- Physical defects (checking for broken type, hickeys*, spots etc)

Hickey - The effect that occurs when a spec of dust or debris (frequently dried ink) adheres to the printing plate and creates a spot or imperfection in the printing.

In many cases the client will inspect and sign off a job on the printing press and indicate acceptable color variances (if they have not already been specified).

POST PRESS CHECKS

Most printing is usually not complete until it is converted into a "finished" product. Post press activities include various types of finishing work such as slitting, embossing, foiling, die-cutting.

Post press checking therefore can include:

Embossing – Checking for defects such as un-sharp edges, pinholes, ruptures and "halos" (shadows around the emboss).

Foil stamping – Defects to be avoided are feathering, color changes, scuffing, peeling and un-sharp edges.

Die-cutting – Defects that are commonly checked for are clean cutting and correct register.

Chapter 4

Proofing

Proofing is a method of producing either hardcopy or electronic facsimiles suitable for viewing a design before printing and for early, intermediate, or final approval.

There are normally two reasons for proofing:

- To make sure that the content is correct i.e. the image is the right way round, all the written information is correct, there are no spelling mistakes, the images are in the right place and do not conflict.
- To ensure that the colors are correct in the design (color proof*)

A color proof is a printed or simulated printed image of each process color using inks, toners or dyes to provide a simulated impression of the final printed reproduction.

In general, proofs will be required at several points throughout the design to print process before the customer signs off a final proof prior to printing.

HISTORY OF PROOFING

Historically press proofing using a rotary or flatbed proofing press was the most commonly adopted method of visualising a design. This method which requires actual printing plates, materials and press time is extremely expensive and time consuming. Although replicating the printing process exactly, this type of wet proof is an inefficient method for initial viewings and for making minor alterations.

Early in the 1970's film based analog proofing

systems emerged, with the development of Cromalin proofing (from DuPont), a toner-based off-press proof. Cromalin proofs are still used today together with dye-sublimation and digital proofing – each of which is described below.

CROMALIN

A Cromalin is an off-press color proof for four color process printing, using separations to construct an image, by the successive exposure and application of adhesive polymer-dry powder toners.

A Cromalin proof is created by exposing a carrier sheet to ultraviolet (UV) light, applying a toner within the cyan, magenta, yellow and black (CMYK) color model and then laminating to a white material.

Cromalins were originally used by the plate-maker to make color adjustments, but are now more widely used by printers, designers and customers as an alternative to mechanical or on-press proofing.

This type of proof will give a close representation of what to expect for color, but it does not replace a press/wet proof. Variations in inks and material substrates used on press will slightly modify the final product (see Figure 4.1).

By the 1980s, film-based analog proofing had largely displaced press proofs. A range of manufacturers besides DuPont further advanced the technology so that matching proofs to press became even more accurate.

In the late 1980's came the digital revolution, during which digital proofing devices began to appear. At this stage dye-sublimation proofing took hold.

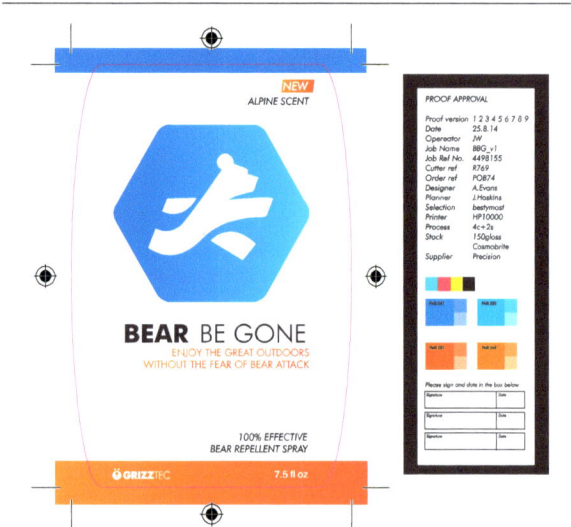

Figure 4.1 - Approval of proofs to agree color with customer – Color calibrated (Cromalin)

DYE-SUBLIMATION PROOFING

Dye-sublimation proofing uses heat and solid dyes to produce photo quality images. The printers lay down color in continuous tones, one color at a time, instead of dots of ink.

Dye-sublimation printers contain a roll of transparent film made up of panels of color. Solid dyes in CMYK are embedded in the film, whilst the print head heating elements vaporise the inks, which then adhere to a specially-coated paper. As the color is absorbed into the paper (rather than sitting on the surface), the output is considered more photo-realistic and more durable than other ink technologies.

By the mid to late 1990's low cost inkjet proofing solutions emerged with both desk-top and large format ink-jet proofing solutions becoming available.

DIGITAL PROOFS

A digital proof is an off-press color proof produced from digital data without the need for color separated films.

Using digital technology the requirement for film to create the image is no longer required. Eliminating the need to make proofs from film was a significant advance for the industry, helping to reduce costs and increase efficiencies.

The job is printed from the digital file using inkjet, color laser, dye sublimation or thermal wax print technologies to give a good approximation of what the final printed piece will look like.

The digital proof is generally less expensive than other proofing methods and can often be produced on the actual face stock of the job thereby adding another element of accuracy.

SUMMARY OF THE ADVANTAGES AND DISADVANTAGES OF DIGITAL PROOFING

The pros and cons of digital proofing are highlighted below;

Advantages

- Accuracy
- True half-tone reproduction
- Consistent results
- Fully automatic
- Easy use and maintenance
- Fast performance

Disadvantages

- Not adequate as a contract proof when metallic inks or phosphorescent inks are involved
- Screening process - does not show possible moiré problems
- Cromalins still have advantages
- Availability of substrates limited using digital proofing

WET PROOFING

A "wet proof" is a proof that is produced using the graphic processes, embellishments and conversion methods used in the "actual" print manufacturing process.

There are no "compromised" processes with this type of proofing and it can be used for all methods of packaging and product decoration. The benefit of wet proofing is that it gives a result that is identical to the actual production run and can be used for sheet-fed, web-fed or direct printing operations.

Equally it can include embellishments and can be converted to a profile shape.

Wet proofing is the most expensive method of proofing.

SOFT PROOFING

With improvements in monitor accuracy and monitor-based proofing software, soft proofing first became viable around 2002.

The term soft proofing is used to describe the technique of previewing a page on a monitor, rather than taking a physical hard copy proof.

The challenge is to achieve accurate color representation on the screen proof that will compare with the actual printed job (bearing in mind that print uses additive color and monitors use subtractive color).

Professional monitors and systems are now usually sold with calibration capabilities to improve the color match between the two. When a monitor at the press and another at a remote customer can be verified simultaneously and display exactly the same output as each other, the need for hard copy proofs is potentially eliminated.See Chapter 6 for a further explanation of soft proofing system developments and their contribution to packaging supply chain developments and collaborative working.

STAGES OF PROOFING

The proof stage is arguably the most critical within the pre-press process, since it is at this point that the customer is provided with a visual of how their final printed piece will look and upon which contractual agreement is made.

A comparison of the process steps involved in producing a wet proof versus a digital proof is featured in Figure 4.2 below. With digital proofing no film output or plate making is required, with digital data directly generating the digital proof.

PROOFING SYSTEMS

The proofing system utilises the press characterisation data to build a color profile which is applied to incoming files and to this profile. If generated correctly, it will allow the resultant proof to closely resemble the final printed piece.

As inkjet printing devices are commonly used to produce these final proofs, the original ink sets need to be manipulated to take into account the ink hue, grey balances, dot gain and overprints found on press.

Stages for full manufacturing wet proof

Design data → Film output → Step and repeat to plate output → Press

Stages for digital proofing

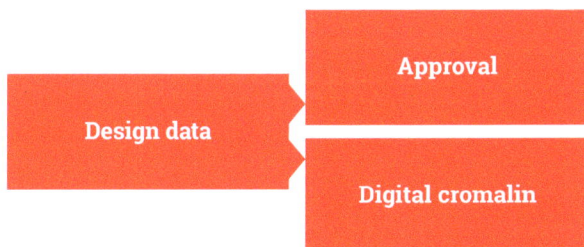

Design data → Approval / Digital cromalin

Figure 4.2 - Stages of proofing – wet proofing versus digital

The incoming file format for these proofs should be in a form that will allow the final inkjet proof to accurately represent the final printed piece. It is for this reason that a Raster proof format is favoured.

A dedicated flexo proofing system using raster data* allows the customer to see how the dot structure and appearance can influence the visual effect of the final print. For example it allows the same image printed at 65 lpi, 100 lpi and 150 lpi to be considered and how much of an impact this screening has on the resultant print.

A raster data structure is based on rectangular or square-based cells

---

The proofing system should also faithfully reproduce other elements that may impact the original design: elements such as rosettes in images and other tonal areas, the appearance of trapping and how this affects color and how colors interact when overprinted.

Accurate spot color reproduction is also sought on the final proof. With the latest ink sets available in devices such as the Epson Stylus Pro series with its additional orange and green inks, this is now achievable for the vast majority of target colors.

With the advances made in inkjet technology, proofs can now be created on multiple substrates such as film and metallic media with the added appeal of printing with a white ink, surface or reverse, to further mimic the final printed piece. A control strip printed automatically on each proof allows the operator to monitor and maintain the output quality.

Apart from the final print, the proof is often the only part of the production process that the client will see. Once the final design is agreed, an accurate color proof may be prepared by ink-jet, dye sublimation, or other suitable procedure. This must be as close to the colors that will be produced on the press as possible.

Only after this approval stage has been completed should the final pre-press procedures be carried out either to the plate making or to a digital press.

SUMMARY

The use of press proofing, analog proofing and expensive proprietary digital proofing systems is on the decline.

Inkjet proofing however, operating with high end Raster image processors (RIPs), has become very accurate and flexible and will continue to satisfy those customers who require a hard proof.

Soft proofing, using color management and software clearly represents the future of proofing in the label sector.

The barriers of high prices of soft proofing systems and more complex associated workflows, however, will need to be overcome before it becomes the industry norm.

Chapter 5

Output – producing the components required for print

The pre-press output function is the part of the manufacturing process that converts the data from the manufacturing specification and the digital design files, to produce the components required for the manufacture of the labels and packaging, such as the printing plates and cylinders.

PRE-PRESS OUTPUT

At this stage the pre-press output procedures includes the creation of layouts, control and application of the step and repeat data, the imaging of films, the preparation and imaging of the printing plate ready for mounting into the printing press (Figure 5.1). Also included in this activity is the accurate mounting of the printing plate and the assembly and checking of the tools and ancillary equipment used for the embellishing and converting processes. All these items are very important factors in the production of high quality print and embellishments and are highlighted in the flow chart below.

PRE-PRESS ELEMENTS

The following list identifies the areas which are important to achieving a fast efficient job change over and press make-ready.

- Accurate agreed technical and manufacturing specification
- Accurate layouts for the step and repeat process
- Imaged film (depending on the process to be used) See Figure 5.2
- Conventional plate imaging system (depending on process to be used)
- Computer to plate (CtP) imaging for the litho, flexography, silk screen and gravure printing processes (dependent on the process to be used)
- Accurately matched inks and ink draw-downs
- Print cylinder specifications

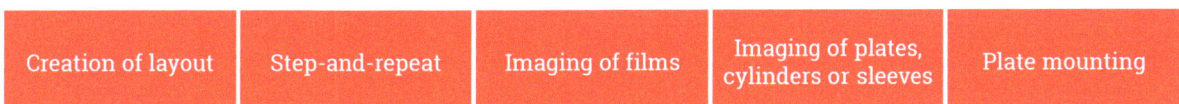

Creation of layout	Step-and-repeat	Imaging of films	Imaging of plates, cylinders or sleeves	Plate mounting

Figure 5.1 - Pre-press output procedures

- Tooling data for embellishing, converting and finishing operations

REASONS FOR LACK OF CONSISTENCY

Within the pre-press arena the following list highlights the main areas that can create production difficulties and therefore additional costs.

- Deviation from the manufacturing specification
- Use of unapproved suppliers
- Image integrity and control
- Changes to repro
- Incorrect die-cutting profile
- Change of specified print processes
- Inaccurate press specification
- Print/embellishment variables
- Color variation
- Substrates characteristics

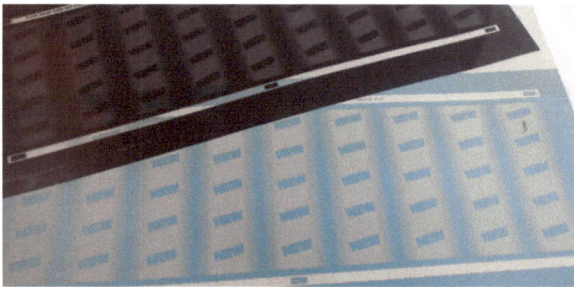

Figure 5.2 - Pre-press output activities include plate-making

PRE-PRESS AND DOT GAIN CONSIDERATIONS

Before imaging the printing plate, the digital files need to be modified to suit the feedback from the press fingerprinting exercise and therefore make allowances for any variations.

Printing processes tend to have a natural condition where the operation is at its most stable for a given set of solid ink densities. This natural condition may not have the desired dot gain or ink trapping requirements to give a good representation of the desired result. This condition must be compensated for (see the section on dot gain and fingerprinting in the previous Chapter 3).

PRE-PRESS LAYOUTS (IMPOSITION)

When the print layout is being planned, due consideration should be given to ensuring that the layout gives optimum production yield and that the amount of substrate that is required for the job is minimised, in order to reduce waste.

Often referred to as imposition, print layout is one of the fundamental steps in the pre-press preparation of a print job. Correct imposition minimises printing time by maximising the number of impressions on a web, thereby reducing cost of press time and materials.

In addition the layout should enable the printing, embellishing and converting operations to be carried out within the parameters of the press and the ancillary equipment being used. For instance consideration should be given to the various inking requirements between one part of a plate and another. If a design calls for heavy solids together with areas of fine detail it may prove expedient to split the color between two printing stations: one to handle the heavy inking and the other to deal with the fine detail. Alternatively this issue can be overcome by introducing an additional screen tone to print underneath the solid area.

Once the optimum layout has been established the step and repeat data can be generated.

The step and repeat file is then used at the film process and plate imaging stage.

STEP-AND-REPEAT

Step-and-repeat is the action of reproducing a number of design images onto film and subsequently the plate, permitting the label design to be printed as more than one label at a time (Figure 5.3). This speeds up production by reducing press running time. In digital workflows the steps-and-repeat designs are imaged directly to plate.

The step and repeat program determines the number of images to be accurately stepped around the printing cylinder circumference and the number of images across the web width.

This is referred to as the number 'across' and the number 'round'. An example would be 7 labels around and 5 labels across yielding 35 labels per revolution of the print cylinder.

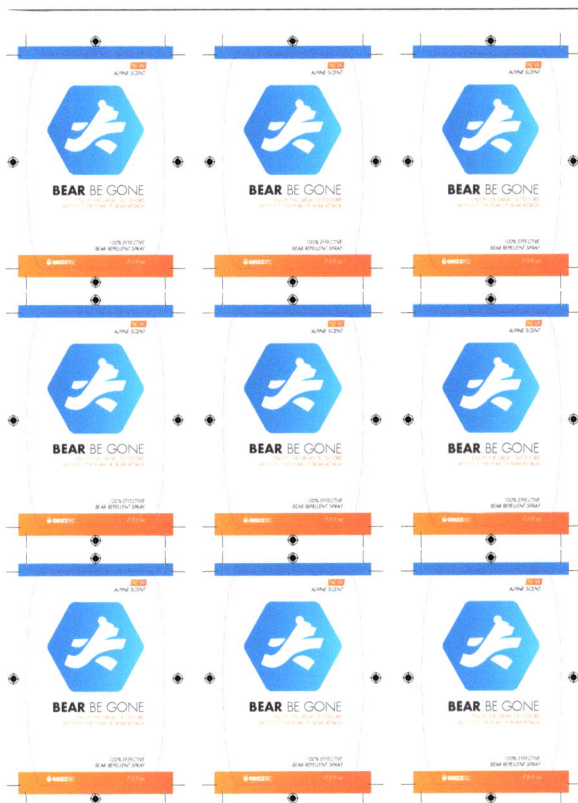

Figure 5.3 - The step and repeat prepares for multi-image printing

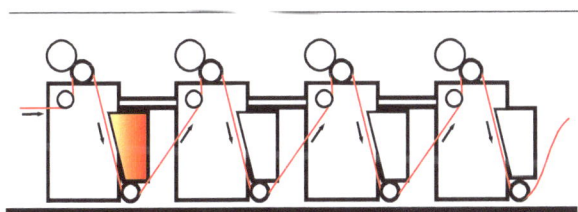

Figure 5.4 - Substrate movement may be caused by heat generated during the printing process

The step and repeat data is used for conventional film imaging and computer-to-plate (CtP) imaging.

ON PRESS CONSIDERATIONS

Modifications to files may need to be made in order to anticipate substrate movement (shrink or stretch) that can occur due to heat, moisture content/humidity (see Figure 5.4)

IMAGING THE PRINTING PLATE

There are two methods used for imaging the printing plate;

- Film based imaging where an imaged film is contacted with the plate and exposed
- The more commonly used CtP (Computer to Plate), where the digital file is imaged directly onto the plate.

COMPUTER TO PLATE (CTP)

With CtP plate imaging a number of the process operations, that are required when using film imaging are removed.

Printing plates are imaged using digital data direct from a raster image processor (RIP) and the image is created using a laser to write the image onto the printing plate, screen mesh or print cylinder. In this way the image distortion caused by the use of film is eliminated and aspects such as dot gain can be controlled more effectively.

The plate making process for each of the major printing systems used in the manufacture of self-adhesive labels is explained below.

LETTERPRESS
Plate imaging

Letterpress plates have a photopolymer layer on a metal or polyester base. The plate making procedure is quick and easy and a plate can be ready for the press within 1 to 1.5 hours depending on plate thickness (see Figure 5.5).

The procedure for most types of plate is as follows:

Exposure to UV light through a matt negative film with a maximum density of 3.5 log density causes the image relief area to polymerise or harden. The log scale refers to the amount of light that can be transmitted through a solid area of exposed film. The lower the amount of light the better the finished plate will be. Wide exposure latitude is provided in order that reversed out images can be exposed, along with normal type and image matter.

Exposure times may vary from one batch of material to the next, therefore it is always advisable to make a test plate when a new batch of plate material is opened. A standard negative should always be kept for this purpose, so that better control can be exercised over the results obtained.

Exposure units may be flat or rotary. If a plate that is to be run on a rotary press is exposed in the flat, due allowance should be made for the distortion that will be created by the curvature around the printing cylinder (known as dispro). The exposure operation is designed to polymerise or harden the areas of the plate that are required to print the image, which is the reason for using negatives for exposure purposes. The image appears as clear film allowing direct access of light to the surface of the unexposed plate.

Wash out

The majority of plate materials are now formulated to permit washing out using plain tap water with few additives. However, for materials requiring special wash-out liquids, the manufacturer's instructions should be carefully followed.

The washing action may be achieved with direct sprays, pads, or brushes. Care must be exercised to ensure that the used wash-out fluid is disposed of correctly and the maker's recommendations followed regarding rejuvenating, recycling, or disposal.

After treatment

This process ensures that no trace of water or wash-out solution remains on the plate. Wiping with a sponge followed by blowing with compressed air is a recommended procedure, ensuring that the still soft surface is not damaged.

Plate drying

In order to rid the plate of any liquid absorbed during the wash-out sequence, it must be thoroughly dried. The working life of the plate will, to a large extent, be determined by the effectiveness of this drying sequence.

Drying times may vary according to the type and thickness of plate material and could (subject to the

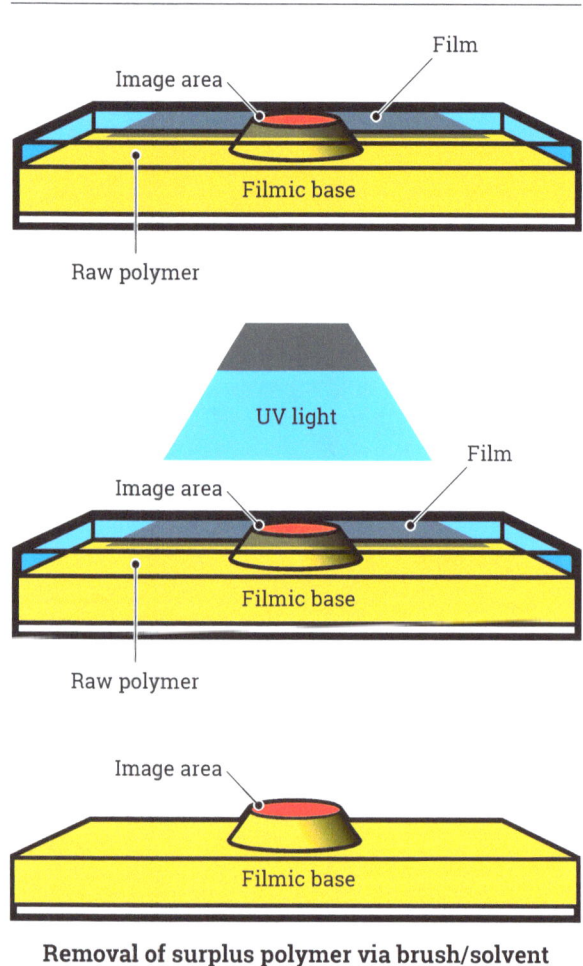

Removal of surplus polymer via brush/solvent

Figure 5.5 - Plate making process illustrating the plate structure, plate exposure and removal of surplus polymer

maker's recommendation), be between 30/80 minutes. It is important that adequate amounts of fresh air are circulated over the plate throughout the drying procedure.

Saturated air will adversely affect the quality of drying and extra time will not necessarily compensate for this. Once fully dry the plate is ready for mounting and use on the label press.

FLEXO
Plates versus sleeves

As with letterpress most flexo plates are based on photopolymer material which is sensitive to UV light. There are however, some flexo plates that are made of elastomer, where the image is engraved by laser and not exposed. Both materials are available in flat or cylindrical form (called 'sleeves'). Sleeves have been successfully used in mid and wide web presses for packaging printing for many years. These systems are now also available for the narrow web presses for label printing.

Plate build up and thickness

The traditional flexo plate has a polyester (PET) carrier film which is very important for the dimensional stability of the plate. The main photopolymer layer is laminated with the PET film. There is a protective film on top of the photopolymer layer called a cover sheet. The film is removed before exposing the plate. This film protects the photopolymer surface from scratches or other damage during transportation or handling.

Within the next few years this plate will disappear from the market. The new generation of flexo plates are direct imaging plates. The structure is the same as a traditional flexo plate, but it has a black ablative layer, also referred to as a LAMS layer (Laser Ablation Mask) on top of the photopolymer layer. (See section on flexo plate imaging below).

Different plate thicknesses are available on the market depending on applications. The plate thicknesses mainly used for label printing are 1.7mm and 1.14mm.

Flexo plate imaging

Two methods of imaging the flexo plate are available:

- Film based imaging
- CtP (Computer to Plate) imaging with a laser.

A negative film is required to image the plate without a black ablative layer. On the surface of the black ablative plate the image is created by ablating or erasing the black layer.

Direct imaging eliminates the need for the negative film and also increases the plate quality e.g. fine negative letters and half tone dots of less than 1% can be imaged on the plate, without any problem.

The general plate-making procedure is similar to letterpress (see Figure 5.5) and involves the following:

BACK EXPOSURE (PRE-EXPOSURE)

The back exposure is carried out from the reverse of the plate with no film in place. This exposure creates a platform base by polymerising the monomer* for better anchoring of printing elements. The thickness of the base has an impact on relief height. The lower the relief height the better the anchoring of fine detail and highlight dots.

---

* Monomer – molecule forming the basic unit for polymer

It is advisable to set a standard for the relief height and it is therefore essential to measure the base thickness of the plate by regularly testing the back exposure.

Main exposure

Whether it is a conventional flexo plate or a direct imaging plate the main exposure is needed for both. This is the exposure to the face of the plate material, through the negative, which forms the image detail that will appear on the finished plate. The details (lines, letters, halftone dots etc.) are reproduced according to the transparent parts of the negative, and cannot be influenced later on. The quality of the relief image subsequently produced depends on the quality of the film negative, the correct exposure time, the condition of the processing equipment and other factors.

The exposure time depends on the type and technical condition of the exposure unit, on the type and age of the tubes providing the exposure light, and on the transparency of the vacuum foil which holds the negative in position during exposure.

It is essential that regular test exposures are carried out using a test negative. These tests are required to check changes in the working of the equipment, in addition to any possible variation from one batch of plate material to another.

Wash out with solvent and drying

Those parts of the plate which have not polymerised are dissolved and removed using a water-based solution or solvent during the wash-out process. The depth of wash-out will vary with the type of plate material being used and the amount of back exposure. The manufacturer's guidelines should be carefully followed.

The wash-out time should be restricted to the minimum in order to avoid undue swelling of the material. Swelling will intensify according to the time that the plate is in contact with the solvent. If the minimum wash-out time is greatly exceeded, fine relief elements of the plate may become detached or distorted.

Solvent wash-out solution has almost disappeared from the market. Water-based solution does not swell the flexo plates in the same way solvent does and therefore the plate drying time is reduced.

The wash-out solvent, which has penetrated into the relief layer of the plate material, is evaporated during the drying process.

Maintaining specific drying times and a uniform temperature across the whole plate in the drying unit is essential. Deviations in thickness may occur if the resting and air drying period before printing is too short.

Wash-out - dry

Another type of flexo plate is available, where the wash-out process is carried out without any solution/liquid at all - this is called thermal processing.

After final exposure the plate is mounted in a dry processing device. The plate is heated and the monomer from the unexposed area brushed-out. This is a very clean and environmentally sound process, which reduces the make ready time for the plate, because no drying is required.

Post-exposure

At this stage the whole plate is exposed without the film in position. Depending on the image (solids, fine lines, dots) the main exposure will allow more or less light to affect the plate. As a result there will be parts of the relief that are less polymerised than others. Post exposure ensures that the finished plate will possess uniform properties over the whole surface.

AFTER TREATMENT

Having passed through these various processes the plate will appear tacky to the touch. If a satisfactory printing image is to be achieved this tackiness must be removed.

The after treatment is designed to remove this tackiness from the relief image by using UV light treatment i.e. exposure under a high energy light source. UV light treatment requires individual exposure times for different plate materials.

High resolution plate-making

Plate imaging and processing for the label industry has undergone numerous changes over the years, from camera, film and plate processing, to the newer technologies using some form of computerised system. Certainly, both computer-to-film and computer-to-plate technologies have had their devotees in recent years for the production of flexo, letterpress and dry-offset printing plates.

High resolution plate-making is a development in the processing of photopolymer plates. A typical laser dot size for high resolution imaging of a flexo CTP plate is around 6 micron (as opposed to 12 micron in the case of standard CTP imaging). Imaging flexo plates at a lineature of 175 to 200 lpi creates extremely fine detail on the plate. With the latest high resolution plate-making systems exposing resolution can now increase to 4000 lpi.

The main high resolution plate-making technologies used in the label industry are;

- Direct imaging on plate by ablating the black layer
- Imaging via laser exposed negative film

Computer-to-film systems utilise UV beams passing through a film mask to change the surface characteristics of a plate surface, prior to a mechanical washout process, drying and plate hardening.

Computer-to-plate systems based on laser

ablation use a laser beam to write an image on to a pre-coated plate surface and in the process destroying the surface coating of the image areas, again enabling mechanical washout, drying and hardening.

Recently new environmentally friendly, computerised plate-making systems have entered the label market. Typical of these new systems is the DigiFlex inkjet C.T.P system which gives excellent quality and high speed plate processing, allowing fast make-readies.

The DigiFlex system is based on a unique ink technology, which gives a very high quality for flexo, letterpress, dry-offset plates and also rotary screens. Creation of the image area on the plate is achieved using a special reactive ink, which is inkjet-printed to produce a UV opaque mask onto a polyester film substrate. The ink chemically reacts instantly with a second reactant on top of the plate to freeze the ink droplets without any time for ink spread. The outcome is a high resolution dot with zero dot gain on the plate and the capability of achieving a 2% dot on the press.

The reactive layer is then transferred from the polyester substrate to the top of the plate using a lamination process. The primer layer, which has no reaction with the plate surface, is washable during the plate development process and can be used with all standard water-washable plates, solvent washable plates, and with rotary screen plates.

After the image has been created on the plate, the rest of the plate-making process remains unchanged, so an already familiar workflow requires minimal adaption.

In Fig 5.6 the DigiFlex plate dots are flatter than dots produced by computer-to-plate laser ablation, enabling easier set-up on the press and better ink capture. Under magnification, the DigiFlex dots are perceived to be sharper and more solid when compared to laser ablation produced dots, which appear more cloudy and grey. Line edges with the DigiFlex system are clearly superior versus laser ablation systems.

Flexo - sleeve build up
The carrier for the imaging surface is a round cylinder or sleeve made of fibre glass. On top there is either a photopolymer layer or an elastomer layer. The cylindrical flexo sleeve has the major advantage that plate mounting is not needed.

Sleeve imaging and processing
In the case of photopolymer sleeves the imaging and processing is the same as for flexo plates, although rather different equipment is needed for plate making.

Elastomer flexo sleeve engraving
This is a direct engraving method, that means imaging and engraving are done in one step by a high powered laser in an engraving machine.

After engraving, the sleeve surface is rinsed with water to remove the remaining burned elastomer. The sleeve is then ready for printing and no further processing steps are required. With the new generation of laser technology it is possible to engrave 200 lines/inch or more.

Another new development in laser engraving technology is available called 3-D engraving. With this technology it is possible not only to change the relief angle and its shape, but also to undercut the highlight dots. With this undercut it is possible to print fine lines, text or highlight dots, with negligible dot gain.

OFFSET LITHO
Plate imaging
With litho plate-making it is important to appreciate the structure of the plate itself in order to fully understand the imaging process. Figure 5.7 overleaf shows the make-up of the litho plate.

The litho printing plate has a totally flat surface and this is referred to as planographic.

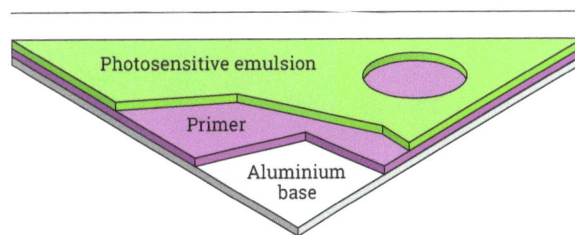

Figure 5.6 - Litho plate structure

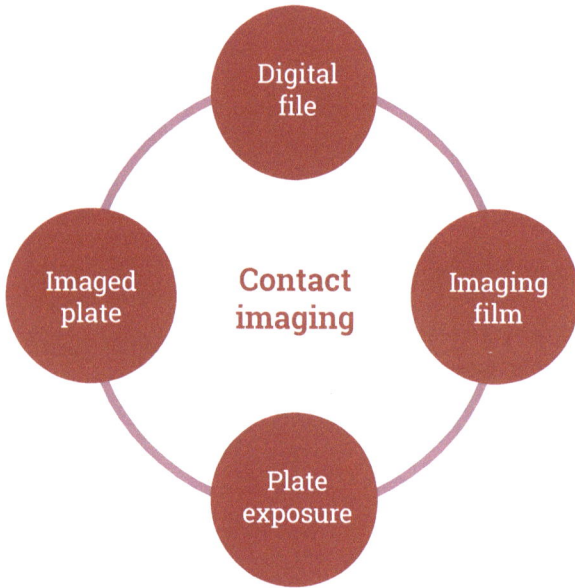

Figure 5.7 - Contact imaging using film

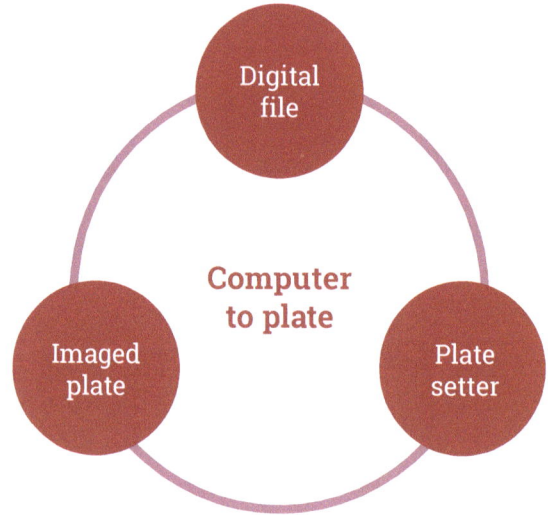

Figure 5.8 - Computer to Plate (CtP) imaging

The modern printing plate is covered with a photo sensitive emulsion. In the illustration below, the aluminum base, the primer layer, and the photo sensitive coating are clearly shown.

Film based imaging

Although most imaging techniques now use CtP, film based imaging is still used. The film negative or positive which is created from the digital file is placed in direct contact with the plate and exposed to a UV light source.

The image from the film is transferred to the printing plates using a photographic process. A measured amount of light is allowed to pass through the film negative thereby exposing the printing plate. On exposure a chemical reaction occurs that activates the ink receptive imaged area. The plate is then developed and the image is chemically fixed. The plate is then ready for positioning into the press. The film based contact imaging is illustrated in Figure 5.7.

Computer to Plate (CtP)

Offset plates are commonly made without the need for film originals. The digital file to be printed is transferred to a CtP device called a plate setter and the image is created using direct laser imaging (see Figure 5.8). After laser imaging the emulsion that remains in the imaged area is removed, leaving it ink receptive.

The CtP imaged plates do not require any chemical processing.

SCREEN
Screen imaging

The screen imaging process works with a screen mesh of nylon or metal strands stretched over a flat or cylindrical frame. The mesh carries a photo sensitive emulsion or coating, which when exposed and processed, washes out the areas to be printed and hardens the coating left behind to become a barrier in the non-printed areas.

Imaging the flatbed screen

An overall photosensitive polymer emulsion coating is applied to the screen material and then dried before a positive imaged film is placed in contact with the flat screen. The screen is then exposed to a UV light source which hardens the emulsion in the non-image areas thus making it insoluble in water. The emulsion in

Polyester screen is stretched tightly
over frame of wood or metal

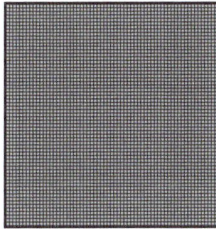

Screen before emulsion

Screen emulsion coated

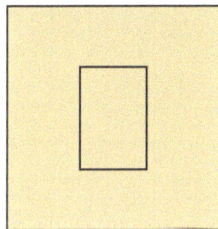

Image created photographically, emulsion
in 'non-image' area is hardened

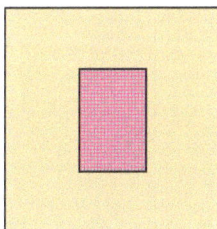

Emulsion in 'image area' is removed by pressure

Figure 5.9 - Stages of flatbed screen imaging

the image area remains soft and the screen is then pressure washed to remove the emulsion from the image areas, before the screen is then dried.

The Stages of flatbed screen imaging

The stages of flatbed screen imaging are illustrated in Figure 5.9.

- Polyester screen mesh is stretched tightly over a frame of wood or metal before it is emulsion coated.
- An image is created photographically on the emulsion and the non-image area is hardened.
- Emulsion in the unhardened image is then removed by pressure washing.

The Squeegee

The function of the squeegee blade in screen printing is to force the ink film through the parts of the screen mesh that forms the printed image.

Print quality will be affected by the type of blade used and the printer will need to select the most suitable blade type. The edge of the blade which is in contact with the screen can be varied in hardness and profile shape and will vary dependent on the type of job being printed. Any damage to the squeegee will affect the print quality.

Rotary screen

The introduction of steel mesh for screen printing led to an important change within the label industry. It allowed the steel screen material to be formed into a cylindrical shape which meant that screens could be fitted into full rotary presses, which are able to run at much higher speeds than the flatbed screen presses. Steel mesh screens can be produced with a wide range of screen value options allowing the printer to choose the most suitable mesh for each job. This choice provides more control over the volume of ink being printed and allows for very high coating weights of ink to be printed, way in excess of the other printing processes.

The rotary screen process adopts exactly the same principles as the flatbed system, but with some key differences. The imaged screen is formed into a cylindrical shape whilst the squeegee blade is placed into the screen cylinder in a fixed position. The screen cylinder which then holds the liquid ink rotates at the

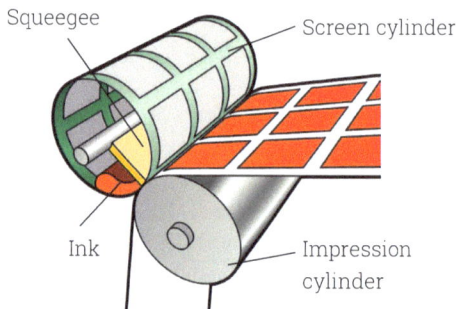

Figure 5.10 - Principle of rotary screen printing showing the ink and squeegee inside the screen cylinder (Illustration courtesy of Gallus Ferd. Rüesch)

same speed as the web being printed and the ink is then forced through the imaged area of the rotary screen and onto the substrate (Figure 5.10).

Rotary screen imaging with film

The imaging of a rotary screen cylinder when using a positive film is very similar to the imaging of a flatbed screen. The rotary screens can be supplied to the printer already made up into the cylindrical shape (Stork system) or can be supplied as flat sheets that are then formed into the rotary cylinder by the printer (Gallus 'Screeny' system).

With all rotary screen systems an end ring has to be fitted into each end of the screen cylinder. This gives the rotary screen the necessary stability and ensures that the screen rotates evenly during the printing operation.

The procedure for imaging rotary screens used in the label industry is as follows :-

1. Metal screen formed into cylinder and end rings fitted.
2. An overall photosensitive polymer emulsion coating is applied to the screen material and then dried.
3. The positive imaged film is accurately positioned in direct contact with the screen and then secured to allow the screen to spin in the exposure unit.
4. The screen plus the secured film is placed into

the exposure unit and exposed to a timed UV light source whilst the screen is rotating.
5. The emulsion in the non-image area is hardened and becomes water resistant.
6. The rotary screen is removed from the exposure unit, the film is removed and the screen placed in the wash out unit, in which the screen is pressure washed to remove the emulsion in the image areas.
7. The screen is removed from the wash-out unit and dried before making ready for the press.

Rotary and flatbed screen CtP imaging

Computer to screen imaging of both flatbed and rotary screens is now widely used. This method of imaging removes the need for film originals and eliminates the exposure process, power washing and drying of the screens.

The digital file which contains the image to be printed is transferred to the imaging unit. A high powered laser then 'burns' the emulsion away, creating the image directly onto the screen. Laser engraving is a digital method of imaging both flatbed and rotary nickel screens. It involves the removal of the emulsion coating in the image areas (i.e. the open areas of the screen). After this the screen requires no further processing and is ready for fitting into the press. In the case of rotary screens, the screen is imaged using a rotary CtP unit, whereas in flatbed imaging the screen remains flat.

CtP imaging reduces the costs associated with the multiple process operation needed when imaging by the traditional method of film contacting. A direct engraved screen produces excellent quality and consistency, with screen resolution of 2540 dpi being produced, to allow fine line work with high contrast to be delivered.

Screens can typically be imaged in 15-20 minutes and because the lengthy drying process is eliminated productivity can be improved, giving a much faster turn round compared to the conventional screen imaging method. Screen material is expensive, but the ability to re-use and re-image screens, especially the rotary screens, has allowed some printers to include a facility which involves stripping off the unwanted image and recoating the screen.

GRAVURE
Gravure cylinder imaging

Gravure cylinders are made of steel and plated with copper. The image area on a gravure cylinder consists of 'cells' that are engraved to differing depths and/or sizes to give the variations in dot size and cell depth. The depth of the cell controls the amount of ink and thereby the strength of color being laid down at a particular part of an image. Very subtle variations in both color strength and fine detail can be achieved.

In the past gravure cylinders were chemically etched, but today the engraving is done using a system which uses a rapidly oscillating diamond tipped stylus. Alternatively the gravure cylinders are imaged using digitally driven lasers to engrave the image. This system removes the problem of inconsistency of the image when a duplicate cylinder is required. With direct digital engraving the image can be simply created and manipulated via computer software.

Worn or obsolete cylinders can be stripped of their image and the base cylinder reused for other printing jobs.

PLATE MOUNTING

For the purposes of this module certain basic principles that need to be followed when mounting plates will be outlined. The complete detail of actual mounting may be obtained from other publications or from the suppliers of equipment designed for the purpose.

The one rule that must be followed in plate mounting is to ensure that the plate will be in the correct position at the first attempt at mounting.

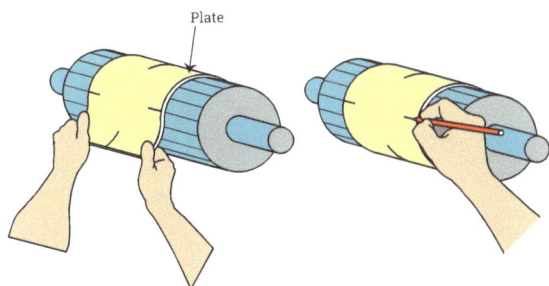

Figure 5.11 - Mounting by hand using guide lines engraved over the surface of the print cylinder

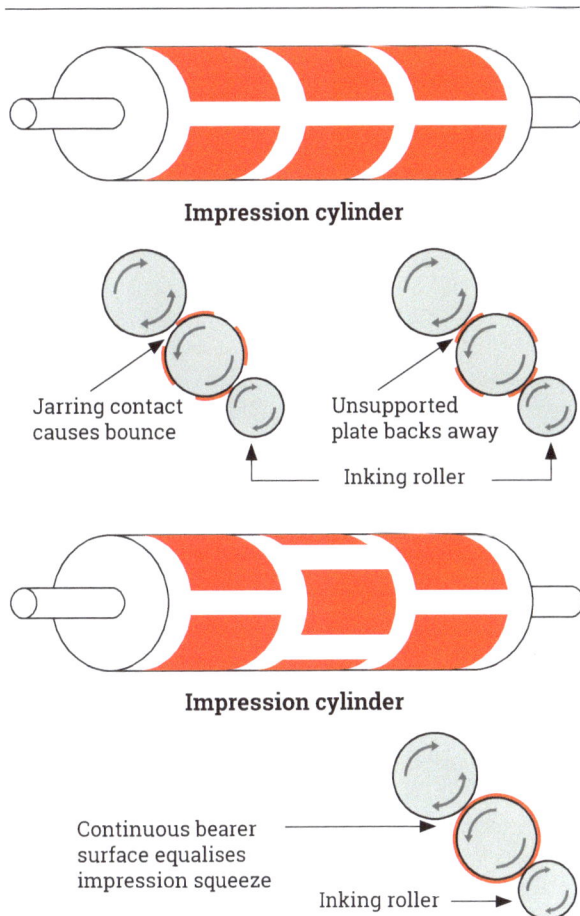

Figure 5.12 - Plate layout for smooth running – incorrect (top) the optimum layout (below)

Figure 5.11 illustrates the correct mounting of a plate by hand using the guidelines engraved on the surface of the print cylinder.

Removal, especially of relatively pliable flexo plates, with the purpose of 'trying again', is very likely to stretch the plate at some point, thus making it almost impossible to place the plate so that it is correctly positioned over the whole of its area.

Figure 5.12 illustrates the optimum method of mounting flexible printing plates around an impression cylinder. The aim should be create a continuous surface that will equalise the pressure across the cylinder. Failure to do this will create an unwanted jarring contact bounce.

A well made plate, correctly mounted on a press which is not regularly maintained, will not provide an accurate printed image. The key areas to check, on a regular basis, in order to achieve accurate printed images throughout a run include bearings, anilox rolls, print cylinders, gears, and of course, general cleanliness. Bearings do wear, in fact wear commences the first time the press is started up - that is the reason why they are fitted in the first place. They are intended as a tightly fitting, but easily replaceable part that can be changed in order to avoid shafts running out of line. In theory they take up stresses and strains caused in parts of the press by heavy tooling, continuous running, heating up etc. The length of life of a bearing will depend on the type of bearing fitted and the amount of stress applied to it. Under a program of planned maintenance all bearings should be cleaned and checked for wear and effective lubrication.

In the flexo, letterpress and litho print processes the print cylinders carry the printing plate and are required to retain even contact with the ink transfer roller and the surface of the substrate being printed or in the case of litho, the offset blanket. There is very little room for deviation from the perfectly true running of the print cylinder. The aim should be for them to revolve within an accuracy of ±0.025 mm.

This will ensure that the pressure applied to both adjacent rolls is reasonably constant. Any measurement, when checking for such tight tolerances, should be made when the press has been running for around thirty minutes. This will ensure that the running parts have warmed up and that any expansion has been taken into consideration.

Double sided tape is used to hold the printing plate to the surface of the printing cylinder (see Figure 5.13).

Some tape manufacturers have a thin layer of foam within their tape which goes some way to smoothing out small deviations in the revolving plate cylinder. With or without such addition these tapes do vary in thickness between one producer and another. This variation in thickness must be allowed for when determining the final diameter of the print cylinder. If this is not done accurate register will not prove

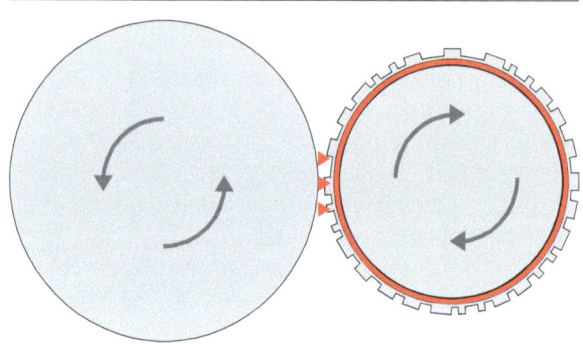

Figure 5.13 - Plate mounting - mounting tape

possible. Also, the same brand of tape should be used on each set of print cylinders, unless the object is to compensate for a slight under or over sized cylinder out of a multi-color set.

Cylinder gears

Gears are required to do more than just transfer power from one point of the press to another. When the gear is manufactured it is intended to be meshed at a certain depth with the gear it is driving. This is called the pitch diameter. When meshing the print cylinder gear with the impression roller this pitch diameter is critical to accurate register (see Figure 5.14). Too deep or too shallow a mesh will cause loss of register between one color and the next. In process color work, the gear for each color should be

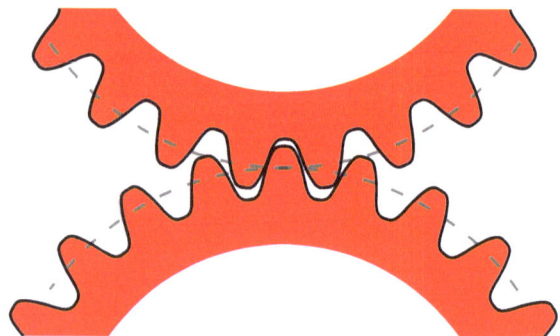

Figure 5.14 - The correct meshing of the gear wheels

set to an identical mesh and checked during a run for deviation, caused through increased temperature or mechanical fault. Regular lubrication is preferred to intermittent applications as this will maintain a constant film of oil and even out temperature fluctuations.

Gearless printing with direct servo drives is state of the art. It eliminates those problems mentioned above and saves time for plate and cylinder mounting.

Print cylinders

When determining the actual print cylinder size on the basis of repeat length in inches, millimeters or number and size of teeth allowance must be made for the thickness of the printing plate and mounting tape and for the effect of pressure and thermal expansion in the cylinder.

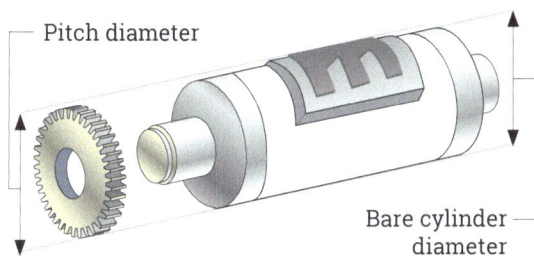

Figure 5.15 - Plate surface and gear pitch relationship

To take an example:

A repeat length of 12 inches (or 96 1/8" teeth) equals an effective circumference of 304.8 mm, but the cylinder itself must have a circumference that is smaller by 3.14 times two thicknesses of plate and mounting tape – say 3.14 x 2 (1.7 + 0.3) mm and a small allowance should be made for the mutually opposing effects of compression and thermal expansion, say 0.01 to 0.03 mm.

Ideally there should be no slack between the print cylinder shaft and gear wheel and also none between the internal diameter of the cylinder bearings and shaft. The whole assembly should fit together snugly. Any movement between these surfaces will upset the

intended pitch diameter and show up as loss of register between one plate and another (Figure 5.15).

On-press controls

A key role of the repro process is to ensure that the press is set–up to the optimum specification and that the correct press settings are maintained throughout the length of the print run.

Although the modern repro process can deliver a highly accurate and consistent reproduction of the original image, factors within each individual printing process and limits on engineering tolerances and wear on the press itself, can adversely affect the desired printed result.

A skilled printer has the facility to make on-press adjustments to vary/improve the printed result to achieve the correct tonal value and color. These adjustments will vary dependant on the printing process being used.

The following paragraphs highlight the areas where on-press settings and adjustments can be made to assist in overcoming any shortfall that may occur in the printed result.

With all the conventional print processes it is possible to make adjustments to the inks being used in order to achieve the correct match. This is called ink mixing and it allows the printer to make fine alterations to the color and also introduce additives to improve the ink performance.

LETTERPRESS

The most effective method of color adjustment on a letterpress machine is the ability to control the volume of ink delivered to the printing plate. The letterpress ink distribution system has an ink reservoir called an ink duct. The ink duct allows the printer to adjust the volume of ink being delivered to the distribution rollers thereby controlling the strength of the color. A reduced volume of ink will give a lighter color and a heavier volume will increase the color strength.

There is very little 'give' in the letterpress printing plate, so in order to achieve a good ink film transfer and a sharp image, the pressure between the printing plates and the impression cylinder or flatbed requires very careful setting.

A skilled operator will vary the hardness of the

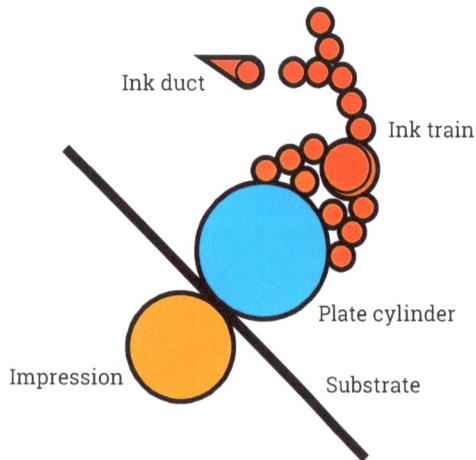

Figure 5.16 - Letterpress print configuration

Figure 5.17 - Flexo Anilox roller. *Source: Mark Andy*

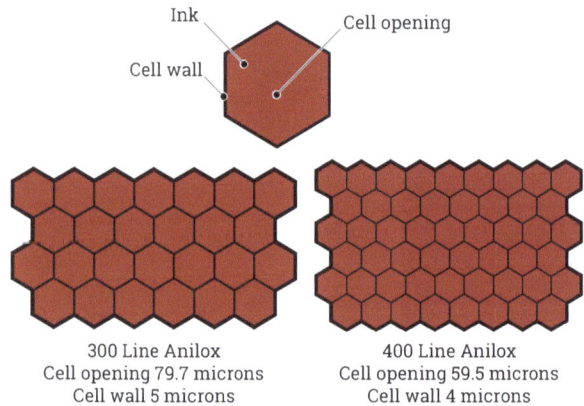

300 Line Anilox
Cell opening 79.7 microns
Cell wall 5 microns

400 Line Anilox
Cell opening 59.5 microns
Cell wall 4 microns

Figure 5.18 - Anilox cell structures

impression roll ensuring that the plate to substrate contact is a 'kiss' touch. Careful balance of pressure between plate and impression roll is crucial as it determines the print quality (Figure 5.16). Too much impression creates a squashed or halo effect and too little impression creates missing dots and poor print.

Accurate inking roller to printing plate settings is very important and adjustments to the printing pressure should be made throughout the press run, to make sure the correct pressure is maintained.

FLEXO

Flexographic anilox rollers in particular are subjected to considerable abrasive wear and careful monitoring of the ink densities and viscosities delivered by the anilox roller is therefore very important (Figure 5.17). It is strongly recommended that each roller is regularly inspected for any damage or reduction in the cell depth.

The depth and shape of the cells is a key factor in the efficient delivery of a uniform ink film to the printing plate and allows the operator to vary the volume of the ink film by changing the anilox roller, to the achieve the correct color The number of cells on an anilox roller are measured in cells per linear inch (CPI) or the cells per centimeter (CPC). As the cell count increases the ink film delivered to the printing plate decreases, (Figure 5.18). It is very important that the printer uses the

anilox roller which delivers the correct ink film, which matches the repro specification.

It is recommended that a record of the ink volume of each anilox should be kept to ensure that the anilox specification is correct for the each print job.

Poor quality flexo printing is usually a result of a soft printing plate, too much impression and a low viscosity ink. Flexo plates are relatively soft compared to the much harder letterpress plate and this soft construction can affect the print quality. A softer plate will transfer the ink film smoother than harder plates, but the likelihood of a squashed effect on the printed dot increases. The squashed dot can easily be identified by a halo effect on the printed dot.

The correct pressures of anilox to plate and plate to substrate must be established at the initial stages of the job make-ready. Incorrect settings will adversely affect the shape and size on the dot.

One of the biggest problems confronting the printer is the problem of dot gain. This effect is created by an 'increase' in the specified printed dot size which affects both the tonal values and therefore the color being printed.

Dot gain can be a result of incorrect plate imaging, but is generally a result of incorrect impression settings or poor engineering or wear on the press.

The printer must ensure that the correct settings between the anilox roll, the printing plate and the substrate are set correctly and maintained throughout the print run.

LITHO

The most effective method of color adjustment on a litho press is the ability to control the volume of ink delivered to the printing plate. The litho press ink distribution system has an ink reservoir called an ink duct. The ink duct allows the printer to adjust the volume of ink being delivered to the distribution rollers thereby controlling the strength of the color. A reduced volume of ink will give a lighter color and a heavier volume will increase the color strength.

In the litho process the inked image is transferred from the plate cylinder onto the offset blanket, which in turn transfers the printed image onto the substrate. The pressure settings between the plate, blanket cylinder and the impression cylinder are important and need to be set correctly. The settings will be dependant on the type and thickness of the substrate being printed (Figure 5.19).

Offset blankets are made of synthetic rubber. The shore hardness of the blanket can vary allowing the printer to choose the correct blanket hardness for a particular substrate. It is important that the release factor of the blanket is correct so that the inked image is fully transferred to the substrate on every revolution of the printing cycle. Ink residue and fibres from the substrate surface can contaminate the blanket and regular cleaning throughout the print run is recommended.

The printer must establish the optimum damping settings during the make-ready process and throughout the print run. The control of the damping

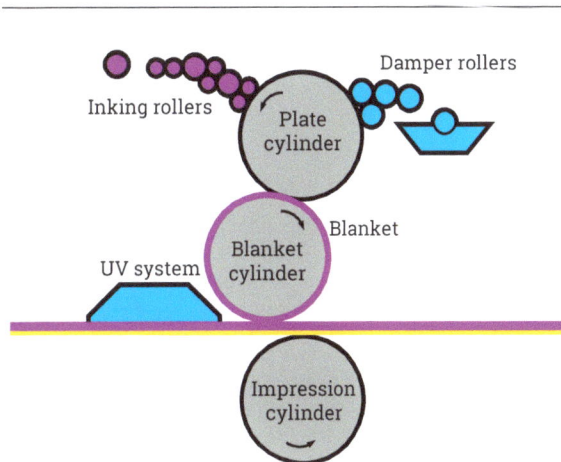

Figure 5.19 - Configuration of a litho printing press

process is critical in achieving the correct print quality. Any imbalance associated with insufficient damping will allow the ink film to contaminate the non-image area and this will result in ink being deposited in the non-image area, thereby creating scumming or catch-up. The problem of plate scumming can be affected by an increase in the ambient temperature around the press and also the heat generated when running the press at speed.

SCREEN PRINTING

The size and depth of the mesh material and the viscosity of the ink are the key areas that control the quality of the screen printed image.

It is most important to select the correct material and mesh count to suit the graphic requirement and the thickness of the ink film that is required. Mesh

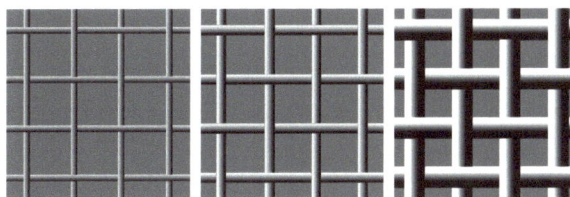

Figure 5.20 - Screen mesh count

selection will impact on the print detail, the stability of the screen and on print registration (Figure 5.20).

The correct viscosity of the ink is also very important in the screen process. The ink must be viscous enough to avoid passing through the screen cells until the exact moment of the printing cycle when the ink is forced through the screen cell by the pressure of the squeegee blade.

Inks with thixotropic properties (this is an ink that thickens up and affects the flow properties of the ink) can offer some advantages to the screen printer, as it becomes fluid when agitation takes place or pressure is applied i.e. via the squeegee blade.

The printer must ensure that the ink is adjusted to the optimum viscosity and this must be maintained throughout the print run. Changes in ink viscosity will affect the color strength and the use of a Zhann cup* measurement is recommended to verify this viscosity.

Zhann cup - dip calibrated viscosity measuring device. A stainless steel cup with a tiny hole drilled in the center of the bottom of the cup. After lifting the cup out of the ink, the user measures the time until it stops flowing to assess its viscosity.

The type of squeegee blade being used can also affect the print quality and the volume of the ink deposit. A soft squeegee will deposit a thicker layer of ink than a harder squeegee and therefore affect the volume of ink being deposited onto the substrate.

The hardness of the squeegee blade is measured in shore hardness and each type is color coded:-

- Yellow has a shore of 55-60 and is suitable for solid work
- Red has a shore of 65-70 and is suitable for solid and line work
- Green has a shore of 70-75 and is suitable for text and line work
- Blue has a shore of 75-80 and is suitable for fine text

The variation in the volume of ink may be small but this can make the difference when matching the exact

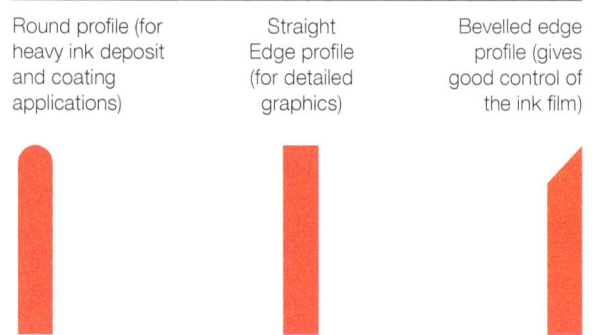

Round profile (for heavy ink deposit and coating applications)

Straight Edge profile (for detailed graphics)

Bevelled edge profile (gives good control of the ink film)

Figure 5.21 - Squeegee blades are available with different profiles

color specification.

The profile of the squeegee can also affect the ink volume being printed and also the print quality, dependant on the content of the image being printed. Squeegee blades are available with different profiles on the leading edge of the blade, i.e. square edge, round edge and bevelled edge being the profiles in common use (Figure 5.21).

A round edge will give a heavier ink volume, but when printing fine definition work the bevelled edge would be more suitable.

GRAVURE

With gravure printing the adjustments that can be made on the press are limited.

As the gravure process prints directly onto the

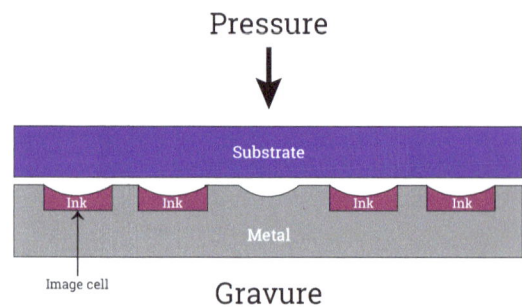

Pressure

Substrate

Ink   Ink   Ink   Ink

Metal

Image cell

Gravure

Figure 5.22 - The gravure printing process

substrate from an engraved cell, in which the size and depth of the cell have already been predetermined, this means that the volume of ink transferred to the substrate cannot be changed i.e. there is no adjustable ink duct, inking rollers or anilox roller used in the gravure press (see Figure 5.22 and 5.23). The two areas that can be adjusted by the gravure printer are the ink viscosity and the angle of attack of the doctor blade. The printer can make adjustments to the ink formulation to achieve the correct color match, but it is most important that the correct ink viscosity is established and maintained throughout the print run. Any variation in the ink viscosity will result in color variation during the print run.

Changes to the thickness and angle of the doctor blade ('wipe') can impact on the ink volume delivered. For example a thicker blade and a shallow wipe will allow a slightly thicker film of ink in the cell, whilst a thinner blade and a steeper angle of wipe will leave slightly less ink in the cell.

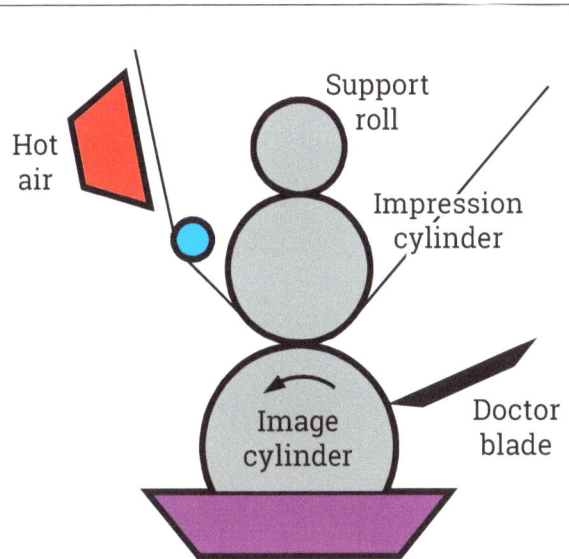

Figure 5.23 - The gravure printing process

SUMMARY OF ON-PRESS SET-UP AND CONTROL FACTORS (BY PRINT PROCESS)

Letterpress
- The volume of the Ink film can be controlled by the printer
- Inking/forme rollers settings must be correct
- The plate to substrate setting needs to be finely set and maintained
- The rubber covering on the impression roller can be varied in shore hardness
- Ink mix formulation can be adjusted by the printer

Flexo
- The anilox roller can be changed to suit the correct volume of ink required
- The anilox to printing plate and the plate to substrate/impression roller has to be finely set and this setting must be maintained throughout the print run
- The ink mix formulation can be adjusted by the printer
- The correct ink viscosity must be established and maintained

Litho
- The correct balance between the plate damping and plate inking must be maintained
- The settings between the plate cylinder and blanket cylinder must be correct for the substrate being printed
- The offset blanket must have the correct release factor for the substrate being printed
- The offset blanket has to be washed regularly and be free from debris
- The volume of the ink film can be controlled by the printer
- Inking/forme rollers settings must be correct
- The ink mix formulation can be adjusted by the printer

Screen Printing
- The correct mesh for the graphics required has to be established before the screen is imaged
- The ink has to be at the correct viscosity
- The squeegee blade has to be the correct hardness/softness for the graphics required
- The type of profile used on the squeegee leading edge will affect the printed result

- The ink mix formulation can be adjusted by the printer

Gravure

- The ink mix formulation can be adjusted by the printer
- The ink viscosity must be correct, as any fluctuation will result in color variation

- Increasing or decreasing the angle of wipe and varying the thickness of the doctor blade will impact on the volume of ink delivered

For more information on the printing processes refer to the Conventional Label Printing Processes module.

Chapter 6

Digital workflow developments and implications for the packaging supply chain

---

With automated and more comprehensive systems, fully digital workflow is increasingly being used in the packaging and labeling sectors.
A number of significant developments have accelerated the move to collaborative working models. These developments and their implications for the packaging supply chain will be explored in this Chapter.

DEVELOPMENT OF NEW FILE FORMATS

The use of the Portable Document Format (PDF) and more recently Job Definition Format (JDF), has provided the printing industry with a predictable and flexible format, that has solved many of the limitations encountered with other file structures.

These new formats are transferable via the internet and can be easily viewed and manipulated, no matter what program has been used to create the file.

The introduction of the PDF and JDF file format has facilitated the development of new and radical supply chain models.

PORTABLE DOCUMENT FORMAT (PDF)

Adobe's Portable Document Format (PDF) has provided the label and printing industry with a solution that enables networked communities to work together. The PDF format allows a document created on one computer to be viewed on any other computer or printed on any device without the receiver having the application that created the document. This offers a unique flexibility which has accelerated developments in digital workflow processes.

JOB DEFINITION FORMAT (JDF)

JDF (Job Definition Format) provides users with the means to describe their jobs electronically.

A JDF is an XML-based file format standard for information exchange, designed to aid communication and automate processes, particularly between designers and the converter.

JDF files can be added to, as job information is gathered throughout the printing supply chain. This level of flexibility is not afforded by a PDF.

EXTENSIBLE METADATA PLATFORM (XMP)

Adobe's Extensible Metadata Platform (XMP) is another file format that is helping in the development of workflow solutions. XMP is a labeling technology that allows data (known as metadata) to be embedded into the file and provides an easy way to deliver project information.

Add to the success of PDF/ JDF, the computerisation of other processes such as pre-flighting, trapping, imposition, digital proofing and digital-to-film, plate or press, and the route to completely digital workflow has now become a reality.

THE DEVELOPMENT OF COLLABORATIVE WORKING

Set against a background of expanding markets and ever more complex supply chains, keeping control of key elements in brand design can be a real headache.

There are however an increasing number of technical solutions that can assist the way product launches are controlled and managed.

The ultimate desire of any brand owner is to achieve the following objectives:

- A harmonised brand where its image appears identical in every country in which it is sold, irrespective of the material it is printed on.
- An increase in the speed of product introductions and brand campaigns whilst at the same time achieving significant cost reductions.

All too often a company's supply chain is characterised by its fragmented and outdated nature, reduced visibility of where the money is being spent and a chronic lack of control, amongst other things. Collaborative digital workflows for visual communications are able to speed up the time to market and reduce the many communication problems and costs surrounding brand development.

Synchronising supply chains (ie integrating purchasing, distribution and marketing) so that brand designs are optimised and harmonised is an effective way to meet customer expectations, whilst at the same time delivering greater control.

COLOR MANAGEMENT SYSTEMS

With advancements in monitor technology it is now possible for brand managers and printers to amend or approve designs instantly on-screen. Calibrated monitors can be located with all the key decision makers in the supply chain. The idea is to guarantee the design and colors they get on the final printed result will be the same as those they have seen and agreed to on-screen (see Soft Proofing Chapter 4).

Waiting for proofs to arrive by post or courier slows down the entire approval process, so proofing on-screen is accelerating the communication across the supply chain and is radically reducing costs and the carbon footprint of organisations.

All members of a team are able to collaborate, to see the file simultaneously, to make comments on it and to approve jobs.

Using state of the art calibrated monitors effectively eliminates all the variables and controls the viewing environment.

The artwork is actually checked against the profile of the printing equipment and the material on which it is going to be printed, so that when the image is rendered on the screen it actually looks the same as it will on the final printed pack or label.

Today a new breed of color management devices and processes are being used to control and manage color throughout the supply chain.

These typically feature the following;

- Customised, high-end monitors
- Color measurement devices
- Kiosk-controlled viewing conditions
- New color science and proprietary methodologies to render RGB into CMYK
- Real time proof suite of web-based, collaborative proofing tools.

BRAND ASSET CONTROL

As we have already seen most of a typical project's costs and time are incurred before it enters production, namely creating, manipulating, approving, translating and finally proofing.

A digital asset library provides a central repository of images that the client can access. This makes it easy for supply chain partners to retrieve previous lines, cross reference brands and mix and match images into a new range, amongst other things.

A lot of time can be spent sending or waiting for images, documents or designs to be delivered by post, courier or email, and making sure that all parties conform to brand guidelines and quality standards.

The development of Open Pre-press Interface (OPI) has supported this activity.

OPI* is a graphics management system that is useful for reducing the time taken to send high resolution files on networks. With OPI systems high resolution images are stored on a central server whilst a low resolution version is supplied to users as a quick and easy way to visualise a job. A key benefit of the OPI system is that when a job is ready to print, the system can reinstate the high resolution files exactly as the user intended.

Workflow protocol developed by Aldus Corporation

The value of brand asset control increases as more and more companies move or expand production overseas.

A vital link in any supply chain is undoubtedly the retailer, many of whom experience problems as a result of poor communication with their suppliers and having to handle an increasingly complex web of relationships. This is why the concept of supply chain management now comes much higher up on the management agenda.

MIS - MANAGEMENT INFORMATION SYSTEMS

Initially the term digital work flow used to describe pre-press workflow, is now being extended to cover the entire process, from the moment a customer asks for a quote or places an order, right through to its delivery and invoicing.

Many software systems can already provide trapping, ripping and imposition tools, but not all are capable of linking such tools with MIS, digital asset management systems, accounting systems and desktop applications, into one comprehensive toolkit. Future workflow systems will increasingly create a synergy between all partners in the label industry supply chain – materials suppliers, converters, distributors, brand owners, etc – and have connectivity to other systems.

All elements of integrated workflow, from design and specification, through all production stages to final distribution and invoicing, working together, will significantly increase throughput and eliminate communication and operator errors in the total label industry supply chain.

Index

www.ingramcontent.com/pod-product-compliance
Lightning Source LLC
Chambersburg PA
CBHW041723210326
41598CB00007B/755